PRACTISE &
PASS
PROFESSIONAL

NUMERACY
TESTS

PRACTISE &
PASS
PROFESSIONAL

ACHIEVE YOUR
PERSONAL BEST

NUMERACY
TESTS

ALAN REDMAN

trotman **t**

Practise & Pass Professional: Numeracy Tests

This edition first published in 2010 by Trotman Publishing, a division of Crimson Publishing Ltd., Westminster House, Kew Road, Richmond, Surrey TW9 2ND

©Trotman, 2010

Author Alan Redman

British Library Cataloguing in Publication Data
A catalogue record for this book is available from the British Library

ISBN 978 1 84455 244 3

Printed and bound by Bell & Bain Ltd., Glasgow

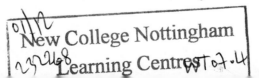

ACKNOWLEDGEMENTS

The practice questions in this book are based on real psychometric tests of numerical ability developed by Criterion Partnership, a leading UK-based test publisher, and used by a wide range of employers. I would like to thank Criterion Partnership for making these materials available to readers of this book.

As someone who struggled at school with maths I'd like to thank my maths teacher, Mr 'Joe' Walder of Dartford Grammar School, who was instrumental in helping me practise and pass my O level in Maths and ultimately ending up working as a business psychologist and developer of numerical tests. I now use statistical techniques regularly and have learned to love the certainty of doing sums.

CONTENTS

✿ CONTENTS

INTRODUCTION

WHAT'S IN THIS BOOK?

Working through all the chapters and practice tests in this book will help you pass numerical tests and answer any questions or anxieties you have about being tested. Here's what lies in store within each chapter.

Chapter 1. Why do you have to take a test anyway?

This chapter sets the scene by providing the background into why numerical tests are used and how employers might use the results, before describing how to get the most from working through the remaining chapters of this book.

Chapter 2. Developing your personal best score

Here we explore the concept of your personal best score and how you can maximise your performance on numerical tests. The chapter describes the range of possible strategies for practising and passing numerical tests, and provides detailed insider information about the mechanics of numerical testing and how to make testing work to your benefit. This chapter is key to your success as it explains the winning approach you need to take to achieve your goal.

Chapter 3. Numeracy practice questions

Don't be tempted to turn to this chapter straight away. You'll achieve maximum benefit from the practice questions in this book only if you work through Chapters 1 and 2 first. This chapter contains some easier numerical practice questions and is a valuable place to begin your practice once you have understood how to develop a successful approach by reading the earlier chapters.

Chapter 4. Numerical reasoning practice questions

We turn up the temperature in this chapter as we introduce more advanced practice questions that measure your numerical reasoning. These questions are best attempted once you have already completed the earlier chapters and the questions in Chapter 3. Along with the right answers, this chapter also provides clear explanations about the reasoning required to arrive at the correct result.

Chapter 5. Numerical critical reasoning practice questions

Numerical critical reasoning questions are the most challenging level of testing used by employers. This chapter contains a series of short tests of your numerical critical reasoning, which even if you're not expecting to carry out in real life, will help you realise your personal best score through the practice and advice the chapter provides.

Chapter 6. Online numerical testing

It is increasingly likely that if you are asked to complete a numerical test you will need to do it online. This chapter provides specific advice about internet-based tests and how you need to adapt the approach and strategies that you would use for paper and pencil testing. This chapter provides detailed insider information about how online tests work, how they are used and why cheats never prosper!

Chapter 7. What happens after taking the test?

Even when the test is over, it's not the time to relax! This chapter explains the way in which test scores are calculated and how the results will be fed back to you. We also explore ways in which you can use your numerical test results to improve your future performance when tested again.

Chapter 8. Your testing rights

As a test taker you should expect to be treated in an appropriate manner by the test user. This chapter describes what your rights are when being tested and what you should expect to happen from the beginning to the end of the testing process. This chapter also contains frequently asked test taker questions to help put your worries to rest.

Chapter 9. Further resources

There is more help out there that you can access to build on the development you achieve by working through this book. This final chapter provides details of test publishers and professional testing bodies that can offer further advice and help.

GETTING THE MOST FROM THIS BOOK

This book will help you enjoy the potential benefits of taking a numerical test by enabling you to practise and pass. To get the most from the book resist the temptation to jump directly to the practice questions. Chapter 2 has important advice for developing your test approach to improve your score.

Chapters 3, 4 and 5 contain practice questions for the three different levels of difficulty that numerical ability tests measure. You should work through these in order regardless of the type of numerical test you will take in future – completing all three chapters of practice questions will help you to maximise your test performance.

CHAPTER 1
WHY DO YOU HAVE TO TAKE A TEST ANYWAY?

Nobody likes taking tests, especially numerical tests. Not even people who work in jobs such as finance or accounting – perhaps they fear being found out as having low levels of numerical ability. The trouble is that numerical tests are increasingly becoming a fact of life. School children are now tested from an early age; our life choices are significantly influenced by the academic assessments we pass, and as soon as we enter the job market psychometric tests appear as an additional obstacle to getting the job we want.

Most jobs these days require us to use some form of numerical ability on a daily basis, whether it's performing mental arithmetic in a bar or restaurant job, reconciling profit and loss reports as a retail manager or analysing complex financial spreadsheets to identify patterns and trends. When ability is important at work, it is likely to be tested to make sure that you are able to respond to the challenges of the role.

For many of us the prospect of taking a test provokes anxiety, worry and even dread. In the days before the test session we fret about our ability to pass, and we get stressed about going through the experience of taking the test and possibly suffer very real physical reactions, such as a racing heart or churning stomach. Numerical tests hold a particular fear for many people, perhaps because of negative experiences at school with complex mathematical concepts such as pi, trigonometry and simultaneous equations. (Hands up anyone who uses these in their adult life. None? Thought not!)

Current graduates are pretty much the most tested generation in history. Not only have they endured a lifetime of academic testing, but they are also the most likely candidates for the jobs that feature the most stringent forms of assessment as part of their recruitment process. With psychometrics becoming increasingly widespread across all types of jobs the rest of us are unlikely to remain entirely untouched by the experience of being tested.

INSIDER INFO

WHAT IS NUMERACY?

While verbal tests measure our degree of literacy, numerical tests focus on numeracy. This is sometimes described as 'number sense'– it is an innate talent that human beings are born with and that develops with learning. Number sense enables us to count, work with different degrees of magnitude (more or less) and is related to our abstract reasoning abilities.

Research by developmental psychologists working with children indicates that number sense is something that infants are born with – it is plumbed in at birth and develops as the child interacts with the world. Even young babies demonstrate number sense in conditions that measure 'object permanence' – the ability to notice when a physical object is removed or increased in number.

Mathematics is distinct from numeracy – which is a reason we sometimes see a difference between numerical test scores and maths exam results. Maths is not just concerned with numbers; it is a system of thought that enables us to work with more abstract concepts such as fractions, negative numbers and precise expressions of magnitude. Maths is a learned ability, while numeracy is an unlearned one. Both are related to each other and both can be developed. But your numerical ability is distinct from your mathematical ability.

The stakes are high when you are being tested by an employer; you need to demonstrate your potential to perform well in the job and differentiate yourself from others in terms of your abilities. You may be competing against other candidates with greater experience of being tested; you may feel that your numerical ability is not your strongest suit.

So there's a lot of fear and loathing mixed up in the whole testing process. But what can you do to develop an effective response? How can you ensure that your performance on a numerical test paints you in your best possible light?

If you work through this book you will develop the skills you need to practise and pass numerical tests. But a good point to start the process is to understand why employers use numerical tests and how numerical tests can help you – there is good news and reassurance to be enjoyed before you begin your journey.

WHY EMPLOYERS USE NUMERICAL TESTS

Test use in the UK is much more widespread than in many other parts of the world and reflects the degree to which the science of psychometrics is dominated by UK expertise. A 2009 survey conducted by the Chartered Institute of Personnel and Development (CIPD) found that of the 755 UK employers who participated in the survey, 40%–50% were using tests of ability. The proportion of this figure using numerical tests is likely to be very high since it is a very commonly tested form of ability.

The figure of 50% mentioned above seems more significant when compared with the proportion of these UK employers who use interviews as part of their selection processes (68%). So while the interview is still the most widely used method of recruitment, the use of ability tests is not too far behind. Here are the results of the survey in more detail:

Type of recruitment tool	Percentage of surveyed companies who use it
Competency-based interviews	69%
Interviews based on contents of a CV/ application form	68%
Panel interviews (more than one interviewer present)	59%
Specific ability tests	**50%**
General ability tests	44%

Type of recruitment tool	Percentage of surveyed companies who use it
Literacy and/or numeracy tests (e.g. basic numerical ability test)	39%
Telephone interviews	38%
Personality questionnaires	35%
Assessment centres	35%
Group exercises (e.g. role-playing)	26%
Pre-interview referencing	19%
Online tests (selection)	17%
Other	6%

For the majority of employers using numerical tests, the main purpose of the test will be for recruitment. Employers have traditionally used numerical tests at the second stage of the recruitment process once they have sifted applications into a shortlist of candidates based on information obtained from curriculum vitae (CVs) and application forms. These shortlisted candidates are invited to visit the employer to take the test, typically on the same day as the interview. The information from the numerical test can then be used alongside the evidence gathered during the interview to make hiring decisions from the shortlisted candidates.

In recent years many employers have moved numerical tests further forward in the recruitment process to the first stage. This is a different approach where the numerical test is used as a 'pre-selection' assessment alongside the CV or application form, that is everybody is tested when they apply for the job, to make decisions about who will be shortlisted and invited for interview. If you are asked to complete a numerical test upfront when first applying for a job the assessment will usually be a web-based numerical test. This means that you don't have to visit the employer to be tested. (Chapter 6 of this book provides specific advice about completing a numerical test online.)

So why do so many employers use tests and numerical ones in particular?

Many employers use numerical tests when selecting people for jobs. They help them to make accurate predictions about whether a person has the numerical aptitudes needed to succeed in a particular job. This information is difficult to assess through interview alone and is critical to performance in a wide range of jobs.

Numerical tests are also used by employers to help identify appropriate training and development for individuals at work. The numerical test is used as a diagnostic and can provide objective information about people's strengths and limitations. This may be used to identify the development needs of the individual or establish their suitability for a training course or programme.

While many employers look for an educational qualification, such as GCSE or A level in Maths, to provide evidence of a candidate's numerical ability, they will still use a numerical test.

INSIDER INFO

PREDICTING JOB PERFORMANCE

Ability tests have been found to be a strong predictor of performance at work compared with other assessment methods. Business psychologists investigate the power of different selection techniques by conducting 'validity studies' that compare scores gathered during a recruitment process with actual performance in the job.

The results of these studies show that there is a much higher correlation between an individual's numerical test score and their subsequent job performance than is the case with other assessment techniques. In general, numerical tests predict job performance better than interviews, CVs and personality questionnaires.

This is because modern numerical tests include tasks and calculations that are the same as those you will use in the job, whereas educational exams tend to include concepts such as geometry, proofs and fractions that are not widely used at work (see page 16).

Numerical ability tests provide information that is less prone to bias and subjective judgement than other forms of assessment, such as interviews and CVs. A numerical ability is a much fairer assessment since it presents all candidates with a level playing field and uses an objective scoring process. Employers therefore include numerical ability tests to increase the fairness and defensibility of their recruitment processes.

HOW NUMERICAL TESTS CAN WORK IN YOUR FAVOUR

Given the widespread use of tests by employers, it is very likely that you will encounter a numerical test at some point in your career. There is some good news though – numerical tests can also be your friend because they offer you several benefits over traditional forms of recruitment such as the interview.

Numerical tests give you an opportunity to demonstrate aptitudes that are important for success. They may help you to show that you are the right person for a job. What's more, you are being given an opportunity to demonstrate your strengths on a level playing field because of the fair and objective nature of numerical tests.

You can use numerical tests to identify areas in which you need to improve your numerical abilities. This can help guide your career progression through the jobs you apply for or the self-development you undertake in your existing role.

INSIDER INFO

CRITICAL NUMERACY SKILLS

Numeracy is the innate form of number sense that we are born with and develop through working with numerical concepts we encounter in life. Numerical tests focus on a range of numeracy skills that are critical to success in a range of jobs:

▶ Arithmetic skills – adding, dividing, subtracting and multiplying
▶ Estimation – being able to conceptualise a number based on some available evidence; how much something would be likely to be
▶ Understanding of basic graphs, charts and other ways of representing numbers graphically
▶ Core calculation skills – such as averages, percentages, ratios, decimals, fractions.

Many of these skills are also fundamental to your mathematical abilities/ knowledge and the level of complexity that you need to demonstrate these skills at will vary with the job. Numerical tests will reflect a significant proportion of these skills in the questions you complete, with high level tests generally drawing on them all at an advanced level.

Numerical aptitudes that are actually used in jobs are tested using questions that are related to work. This means that you can use the experience of completing the numerical test to help you ensure that you pick a job that actually will suit you, that you will do well in, and that you will gain satisfaction from.

CHAPTER 2
DEVELOPING YOUR PERSONAL BEST SCORE

Testing is nearly always high stakes; if you've been asked to take a numerical test it's generally to demonstrate that you have the right abilities for a job or training programme. When people fail tests it's often hard for them to understand how they could have performed at a level that would have got them the job or placement they wanted.

Sometimes people blame their luck or mood on the day of the test. When feeding back low test scores to others they will often say that their poor performance was the result of not being on top form on the day, or simply being unlucky with some of the answers they guessed.

Other people will blame the test itself – that the result is wrong and not a true reflection of their real numerical ability. Yet others explain poor test results as resulting from nerves, inexperience, confusion or ill-health. Only a few people attribute their low numerical test score to their low numerical ability.

In reality low test scores can be a combination of all these kinds of factors. And the good news is that you can do something about it.

WHERE DO LOW SCORES COME FROM?

Business psychologists work with people to help them to develop their performance on tests, and you can use these techniques to raise your own performance. Two people I have worked with in the past illustrate the different practise and pass approaches that are available to you.

Claire was a recent economics graduate who had worked part-time with a supermarket chain during her degree course. After graduating with good results Claire applied to the supermarket's graduate scheme to join the firm as a full-time employee in the finance department of its head office.

Part of the first round of the recruitment process was an online numerical critical reasoning test. On her initial application Claire failed to score at a level that formed the pass-mark for invitation to the assessment centre.

Undeterred, Claire worked hard to improve her score for further job applications in the subsequent months. She practised by completing lots of degree-level maths questions from past papers from her degree to try to improve her performance. But in her second and third applications to graduate employers her numerical test results still let her down.

Ali was a bank manager with many years' experience, having run several branches across the south-west of the UK. Despite being successful, Ali wanted to change careers and applied for a place on a part-time degree course in research and analysis at a prestigious university, which was sponsored by a large insurance company. The degree would qualify Ali to work as a risk analyst within the company, substantially broadening his horizons and earning potential.

An examination for the university course included a numerical reasoning test, which was used to decide who would be invited to attend an interview with the course admissions tutor.

Ali felt nervous about the numerical test despite his job as a bank manager and hoped his experience in working with finance would benefit his performance. He was therefore disappointed to learn that his test score failed to meet the entry requirements for the course and he would therefore not be invited to interview for a place that year.

Both Claire and Ali were understandably frustrated; Ali because he felt that his numerical test score did not reflect his experience and real abilities and Claire because of her test score's stubborn refusal to improve despite all of the practice she had put in over the past six months.

In helping Claire and Ali to practise their numerical tests I encouraged them to refocus their efforts on **achieving their personal best** numerical test scores.

WHAT IS YOUR PERSONAL BEST?

Your personal best score is the maximum you can hope to achieve on a specific type of test. To achieve your personal best score you must tackle all of the causes of low test scores. These causes fall into three broad categories:

▶ Lack of knowledge
▶ Lack of strategy
▶ Lack of practice

Ali had a relatively high level of numerical ability but this was not reflected in his test score. He failed to score his personal best numerical test result because of a lack of knowledge and strategy. This led him to be a poor test taker; he lacked understanding of how the test worked and the best approach to take during the test session. His well-practised and developed numerical ability was let down by his inexperience and poor test-taking strategy.

Claire completed a lot of numerical practice questions, which improved her test score a little. She still failed to achieve her personal best score as a result of poor test-taking strategy and a narrow range of practice.

To help Claire and Ali achieve their personal best numerical test scores I asked them to address the causes of their low scores by making improvements in the areas that were holding them back. As a result Ali went on to win a place on a university course and Claire made it through to the interview for a similar job with another retailer and ultimately landed a graduate job.

INSIDER INFO

THE SCIENCE OF PSYCHOMETRICS – YOUR PERSONAL BEST SCORE

The term 'personal best score' is drawn from classical test theory. This is a model used by business psychologists in the development of ability tests to interpret candidates' results. Classical test theory states that for any given test people have a true score of ability. This true score perfectly represents the individual's ability on that test.

For example, you have a true score that represents your numerical ability. The challenge for test developers is to ensure that the numerical ability test accurately measures your numerical true score. Numerical tests that are badly designed or poorly administered introduce error into the measurement of your true score. This distorts the measurement of your numerical ability; the test no longer accurately measures the real level of your ability – your personal best.

The challenge for you as a test taker is to make sure that you demonstrate your true score when you take the test. If you are ill-prepared, unpractised or have poor test-taking strategies you will not demonstrate your true score no matter how well the test is designed.

You can use the same techniques to practise and pass numerical tests by following the three steps described below for developing your personal best score.

THE THREE STEPS FOR DEVELOPING YOUR PERSONAL BEST

Achieving your personal best score means addressing all of the possible causes of a low test score.

1. You must increase your **knowledge** of the test so that you understand how it works.

2. You must develop your test-taking **strategies** so that you are not let down by a poor performance on the day.

3. You must **practise** to develop your numerical ability to its maximum possible level.

STEP 1: KNOWLEDGE – KNOW YOUR ENEMY!

The first step to achieving your personal best score on a numerical test is to understand what a numerical reasoning test is and how it works. If you are unprepared in terms of your knowledge of the test you are unlikely to demonstrate your maximum level of ability.

This section will help you to develop your knowledge of numerical tests and enable you to arrive at the test session well prepared and ready to demonstrate your best. You can take your preparation further by finding out more about the specific numerical test you are due to complete.

See 'Find out more about your specific test' later in this chapter for more advice.

Numerical tests and the demands of the job

Employers use numerical ability tests because these tests provide information about your aptitude or potential for performing well in certain critical demands of the job. You use your numerical ability to perform effectively at a range of tasks such as:

▶ Using mental arithmetic to keep totals, add up prices and calculate discounts
▶ Learning from tables and graphs in training materials
▶ Analysing complex numerical data

- Understanding financial results, profit and loss accounts and accounting reports
- Solving complex calculations
- Applying rules, procedures and processes correctly
- Making decisions based on numerical data
- Applying abstract reasoning or lateral thinking
- Solving problems that require a number of systematic, separate steps
- Making estimations based on existing numerical data
- Calculating percentages, averages and ratios.

Numerical tests measure abilities that are central to much of what people do in the vast majority of jobs, which is why they are one of the most commonly used types of tests. You can therefore gauge the level of demands within the job from the difficulty of the test.

Numerical tests are often used alongside other types of test within a recruitment or training process, such as a verbal test. This can be a reflection of the wide range of abilities required for effective performance in a job or sometimes employers are interested in understanding your general ability by asking you to complete more than one type of test. If you asked to complete a numerical test as part of a job application it is very likely that you will also be asked to complete an additional test of ability; the most likely choice is a verbal test. Remember that you need to focus time on developing your verbal score as well as your numerical score; you could do this using *Practise & Pass Professional: Verbal Reasoning Tests* that is also part of this series.

How numerical tests work

Modern numerical tests used as part of recruitment and training processes measure your ability to understand, manipulate and make decisions based on numerical information.

INSIDER INFO

NUMERICAL ABILITY TESTS AND MATHS EXAMS

It's not uncommon for the results of numerical ability tests to be at odds with results from mathematics exams such as GCSEs and A levels. Even Maths graduates can score badly on a numerical ability test. Why is this?

One reason is that tests and exams are two different kinds of measurement. Exams measure knowledge – you learn a syllabus and then you're tested on it. In the case of a maths exam this will often be a range of different mathematical techniques such as algebra, geometry, probabilities and so on. To perform well on the exam requires a range of attributes including good recall, good exam technique and an effective approach to revision. Exam results can also be affected by the underlying motivation of the student – people who try harder can beat others with higher levels of raw ability by simply putting in more hours.

Numerical tests do not require you to learn and recall a syllabus of knowledge – all those candidates whose exam results were lifted through hard work and good revision technique are placed on a more level playing field where raw numerical ability is the critical differentiating factor (although this can be enhanced through practice of test-taking techniques).

Another reason the results from numerical tests can vary from exam results is because of the difference between mathematics and numeracy.

Numerical tests require you to understand some numerical information and then answer a series of questions relating to the information. Good quality numerical tests keep the use of words to a bare minimum so that you are not required to apply too much verbal ability to do well in the test. The numerical information can be presented in many forms.

Tables of information

Tables can be used to present numerical data such as prices, results and costs. To make sense of the data you need to interpret what each row and column tells you. Here's an example:

INSIDER INFO

KEEPING IT PURE: TEST INTER-CORRELATIONS

Test developers aim to design numerical tests that are as pure and accurate a measure of your numerical ability as possible. This means maximising the use of numbers and minimising the extent of other information such as words, diagrams or pictures.

Numerical tests that contain too many words, or other types of non-numerical content, run the risk of contaminating the assessment that is being made of your numerical ability. Numerical test questions with lots of words, for example, place demands on your verbal ability as well as your numerical ability. So in order to arrive at the correct answer you need to employ a combination of abilities – in this case verbal and numerical.

It is very difficult to design a numerical test with no words or other type of non-numerical content. And indeed, any numerical test that was completely free of language would not reflect real-life materials from the workplace. However, modern occupational tests reduce the demands placed on your non-numerical abilities by keeping words to a minimum and presenting the numerical information in tables, charts and other figures. In this way your test performance is based on a pure measurement of your numerical ability.

Candidate's name	Numerical test score	Interview score
Claire Stevens	50	3
Susan Hill	35	7
Ali McGyver	20	2
Nick Driver	64	6

The test question related to the above table will require you to interpret the data in it to pick the correct answer. Here's an example of a very straightforward question based on the table of data above:

'What was Ali McGyver's numerical test score?'

The correct answer to that question is **20** – to arrive at the answer you needed to interpret the table correctly by reading the values in the 'Numerical test score' column in the row for Ali McGyver's results.

A more complex question is:

'What candidate has the biggest difference between their test score and interview result?'

The correct answer to that question is **Nick Driver** – to arrive at the answer you needed to interpret the table correctly by reading the values in the 'Numerical test score' and 'Interview score' columns and subtract one from the other for each candidate. When you subtract the interview score from the numerical test result for each candidate you arrive at the following values:

- ▶ Claire Stevens – a difference of 47
- ▶ Susan Hill – a difference of 28
- ▶ Ali McGyver – a difference of 18
- ▶ Nick Driver – **a difference of 58**

This question therefore requires you to demonstrate more advanced numerical ability by interpreting the data in the table correctly and then using a systematic problem-solving approach to calculate the correct answer.

Graphs

Numerical data can be presented in a graphical format, and graphs are often used to present summary information. To make sense of the data you need to understand what each element of the graph tells you by correctly interpreting the information on the vertical and horizontal scales. An example is given below.

CANDIDATES TEST SCORES & INTERVIEW RESULTS

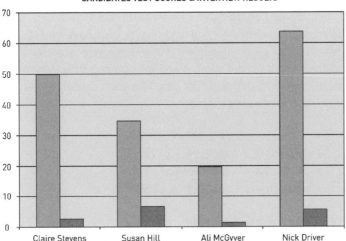

The graph above presents the same data as the table in the previous example. The questions based on this numerical information could follow the same format, such as the examples below.

'What was Susan Hill's numerical test score?'

The correct answer to that question is **35** – to arrive at the answer you needed to interpret the graph correctly by reading the value in the 'Numerical test score' column for Susan Hill. It's quite a demanding question because you have to estimate the value of the test score since it does not fall on a precise value on the test score axis.

Another question could be:

'What candidate has the smallest difference between their test score and interview result?'

The correct answer to that question is **Ali McGyver** – to arrive at the answer you needed to interpret the graph correctly by reading the values in the 'Numerical test score' and 'Interview score' columns for each candidate. You could choose to arrive at the answer by judging which candidate had the smallest size difference between their two columns, or you could read off the values for each candidate's test and interview scores and subtract one from the other. When you subtract the interview score from the numerical test result for each candidate you arrive at the following values:

- ▶ Claire Stevens – a difference of 47
- ▶ Susan Hill – a difference of 28
- ▶ Ali McGyver – **a difference of 18**
- ▶ Nick Driver – a difference of 58

You can also see these differences on the graph in a more visual manner without the need for calculations – this would be the quickest way to answer the question if you were under pressure for time, but it may not be the most precise.

This question therefore requires you to demonstrate more advanced numerical ability by interpreting the table of data correctly and performing a systematic problem-solving approach to calculate the correct answer.

There are many types of graph: the example above is called a column graph and is probably the most familiar type, but we could have used other types such as bar graphs (horizontal columns), line graphs or scatter graphs. All types apply the same principles of having a horizontal (x) axis and a vertical (y) axis, which you use to interpret the data. It is unlikely that you will encounter more complicated graphs, such as one with more than two axes, in a modern numerical test.

Pie charts

Strictly speaking these are a type of graph, but they are worth treating differently because they are used in tests in a slightly different way to the types of graph described above.

The main value of pie charts is in presenting information about proportions or percentages. Pie charts show us how a total is shared among a number of separate categories – like cutting a whole pie into different slices to share among different members of a family. These slices can be equal or different (perhaps to reflect the different family members' appetites or dietary obsessions). Here's a more work-like example:

ATTITUDES TOWARDS NUMERICAL TESTS

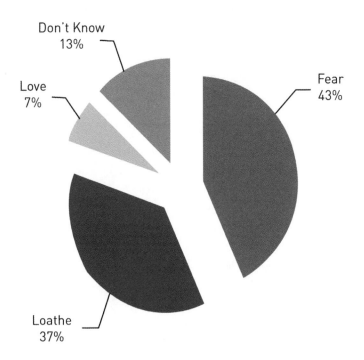

Don't Know
13%

Love
7%

Fear
43%

Loathe
37%

A pie chart requires you to find the data being described in the question. A relatively simple example would be:

What percentage of people fear and loathe numerical tests?'

The correct answer would be **80%** – to arrive at the answer you needed to interpret the pie chart correctly by reading the percentage values for 'fear' and 'loathe' and adding them together: 43% + 37% = 80%.

A more advanced question would introduce extra numerical information, which you would need to use in conjunction with the pie chart:

If the numerical test attitude survey interviewed 1,400 people how many of them love numerical tests?'

The correct answer would be **98** – to arrive at the answer you needed to interpret the pie chart correctly by reading the value for the percentage of people who said that they love numerical tests. You then need apply this figure to the new piece of information–that 1,400 people were surveyed. You then need to calculate what 7% of 1,400 equals:

▸ 1,400 ÷ 100 = 14 (i.e. 1% is 14 people)
▸ 14 × 7 = 98

Therefore the answer is 98 people.

Other formats for presenting numerical data

Tests can use other variations of format to present data, but they will fall into one of the three broad categories of tables, graphs or pie charts. Despite their variations of format though, one thing numerical tests tend to have in common is the use of the multiple-choice format for their questions.

Multiple-choice answers

Numerical tests usually adopt a multiple-choice format for answering the questions. The convention for multiple-choice questions is to have five options from which you must pick the answer you think is correct.

Here an example based on the same numerical information above.

1	If the numerical test attitude survey interviewed 1,400 people how many of them love numerical tests?			
A	B	C	D	E
7	14	52	98	100

We know from the previous example the correct answer is **D: 98**. Note how some of the answers are quite close to the correct answer. That is, they could be the answers you'd get if you made mistakes in the calculation or interpretation of the numerical data.

Some tests may provide more than five multiple-choice options to try to make guessing more difficult, but the most common practice is to use five.

INSIDER INFO

DISTRACTERS: WILY TEST DEVELOPERS TRY TO CATCH YOU OUT

We test developers do like a good Heffalump trap and we design our multiple-choice options to make sure the correct answer is well hidden and to catch out the unwary candidate.

The wrong answers in a multiple-choice question are called 'distracters'. Their job is to prevent the correct answer from being immediately obvious to the test taker by presenting some plausible alternatives. Some distracters will be more plausible than others; one or two will be fairly close to the correct answer, whereas others will be less likely to be mistaken as the right answer.

Distracters are often generated during the development and trialling of the numerical test. The test developer will collect all the common wrong answers that trial candidates come up with when attempting the pilot version of the test (which is open-ended rather than multiple-choice). Distracters are therefore based on the common mistakes that people make when answering each question – which is why they can be so distracting.

Older question formats

If you are asked to complete an older numerical ability test you may find that the question format differs from the multiple-choice approach used in modern tests. These more traditional formats are based on tests used in education. One example is to have an open-ended approach, where you are not given a multiple choice format to choose from. Instead you have to write your answer (and sometimes your method of working out the answer) in the space provided. An example is given below.

Candidate's name	Numerical test score	Interview score
Claire Stevens	50	3
Susan Hill	35	7
Ali McGyver	20	2
Nick Driver	64	6

'What is the average test score for this group of candidates?'

(Write your answer along with how you worked it out below)

The correct answer is 42.25, based on the following calculation of the average test score:

- ▶ 50 + 35 + 20 + 64 = 169
- ▶ 169 ÷ 4 (the number of candidates) = **42.25** – this is the average test score

Another old-style approach is the numerical problem-based approach. With this type of question there are no tables or charts of numerical information. Instead, each question contains a separate numerical problem. An example is given below.

'A survey, in which a large number of people were interviewed, was conducted to understand the range of attitudes towards numerical tests. The results showed that of the people interviewed, 43% experienced fear in relation to numerical tests; 37% experienced loathing; 7% loved numerical tests and 13% were in the 'don't know' category. If the survey interviewed a total of 1,400 people how many of them love numerical tests?'

One of the key problems with this type of approach is that it draws too heavily on the test taker's verbal ability. Before you can even begin to solve the numerical problem you have to have a certain level of verbal reasoning to make sense of the question. This creates two issues: firstly, the test is no longer a pure measure of your numerical ability because the assessment is contaminated with verbal ability; and secondly, it is potentiality unfair to candidates whose verbal abilities are constrained in some way – those who speak English as a second language for example. This would mean that a non-native English speaker who had high levels of numerical ability would be likely to receive a lower score because of the strong verbal content of the test.

For these reasons this format of numerical test is unlikely to be found outside the classroom. This approach does not tend to appear in modern, occupational tests of numerical ability and if you are given a test that uses this format it is likely that the employer has chosen to use an out-of-date test.

The good news for you as a test taker is that the approach that you can take to practise and pass modern numerical ability tests will work for this old-fashioned format as well. The problem-solving processes you use to arrive at the correct answer to the question above is the same as the approach you'd use for correctly answering a normal numerical test question (once you've got past the additional verbal complexity). This approach is described in the chapters that contain the practice questions in this book.

In fact, the problem-based question above is asking the same question as the previous example – only without using the table of numerical information; the answer is once again 98 people. If you are given an old-style numerical test that contains this type of wordy problem-based questions, the work you would have done in this book to practise and pass tests will hold you in good stead.

You can also prepare yourself further by making sure that your verbal abilities are up to the task of deciphering this type of question. One way of doing this is to also develop your approach to taking verbal tests. *Practise & Pass Professional: Verbal Reasoning Tests* is a useful tool for your development.

Some even older numerical question formats

Very traditional numerical tests sometimes use the formats described below.

Number sequences

This type of question requires you to identify a missing number in a sequence of numbers, such as the example below:

Which number comes next in this sequence:

1, 2, 3, 5, 8, 13, 21...

While this format minimises the use of words and therefore avoids the problem of contamination by verbal ability it is still not a pure measure of your numerical ability; instead it draws on your sequential reasoning ability – the ability to reason with sequences.

The answer in the question above is 34 – each number is the sum of the two numbers that precede it (e.g. $1 + 2 = 3$, $3 + 2 = 5$, $5 + 3 = 8$, $8 + 5 = 13$, $13 + 8 = 21$ and finally $13 + 21 = 34$). Your performance on this type of numerical test will definitely benefit

from any work you've done to develop your numerical ability –
but you will need to draw on your sequential ability as well.

Fortunately this format of question is restricted to educational
tests and very old ability tests – neither are very likely to be
used by a modern employer.

Diagrammatic questions

This type of question uses shapes or other forms of diagram
containing numbers. An example is given below.

'What is the missing number in the fourth diamond?'

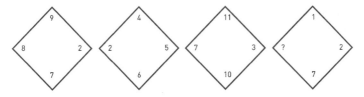

The correct answer to this diagrammatic numerical test
question is **4**. Once again it requires you to draw on additional
abilities such as sequential and diagrammatic abilities to make
sense of the question in order to solve the numerical element
at its heart.

In this case the sequence of numbers works by adding together
the numbers in the top and bottom corners of each diamond
and then dividing that sum by the number in the right hand
corner. So the missing figure in the fourth diamond is 1 + 7 =
8 ÷ 2 = 4.

Again, if you see this question format you can make the
assumption that the employer has chosen to use a more old-
fashioned test, which might reflect their more general style
as an employer. In terms of practising and passing numerical
tests, the work that you complete within this book will help you
develop the underlying numerical ability that you will need to
answer this old-style question format correctly.

There are other types of numerical test and question format that are not used in contemporary numerical test design. However, with all of these you can still draw on the numerical abilities and test-taking strategies that you develop in this book – but you are unlikely to encounter them in the modern workplace.

INSIDER INFO

NUMERICAL TESTS AND 'CANNOT SAY'

More advanced numerical tests may include the 'cannot-say' option in the multiple-choice distracters. The cannot-say answer is a response option we usually associate with verbal reasoning tests, where you have to choose between 'true', 'false' and 'cannot say' when answering questions. In the context of a verbal test, the cannot-say option is used when it is not possible, using the information presented, to say whether a statement is true or false.

The cannot-say option in numerical tests is somewhat different. Unlike verbal reasoning tests it will not appear as an option for every question. Instead it appears in questions where it is either the correct answer or a distracter to the correct answer. If some questions in a numerical test feature a 'cannot-say' option then it is a fair bet to assume that at least one question in the test will have cannot-say as its correct answer. However, you must take care not to misuse the option since this will adversely affect your test score.

The cannot-day option is the correct answer to questions that:

▶ Have more than one possible correct answer
▶ Contain insufficient information to answer the question (i.e. you would need additional data not given).

You should only select the cannot-say option when one of these conditions is fulfilled. You should not use the cannot-say option as a general 'don't know' response for questions that you find tricky. Instead you should focus on the other options and try to narrow down the most likely to be correct.

Timing

Numerical tests usually have a strict time limit within which you must answer the questions. The questions that you answer within this time form the basis of your score for the test, which is typically based on a calculation that compares your number of correct answers with an average for that test. Some tests are designed to give you plenty of time to answer all of the questions while others use a very tight time limit, which means you are unlikely to answer all the questions before the test ends. In both cases you should work quickly but accurately.

▶ See the 'Step 2: Strategies' section later in this chapter for more information about how tests are timed.
▶ See Chapter 3 for more advice about how to answer numerical test questions correctly.
▶ See Chapter 7 for more detail on the way in which tests are scored and what the results mean.

Different levels of numerical testing

Employers should ask you to complete a numerical test that reflects the demands of the job and the level of difficulty of those demands. Numerical tests are pitched at three levels of difficulty:

1. Numeracy

2. Numerical reasoning

3. Numerical critical reasoning

This book focuses mostly on levels 2 and 3 because these are the most frequently tested types of numerical ability. However, you will still find it valuable to work through the level 1 practice questions in the book because these can help you to polish the foundations of your numerical test performance.

Numeracy tests

Numeracy tests are the simplest form of numerical ability test. These tests are based on clear and straightforward numerical information and require you to demonstrate your ability to interpret the numbers correctly and perform simple calculations.

Numeracy tests tend to use the multiple-choice format. You will generally be allowed to use a calculator and rough paper to help with any working out. This may not be the case if the employer is interested in discovering your level of mental arithmetic – for this type of basic numeracy test you would be expected to perform any interpretation and calculations in your head.

In most contemporary jobs you do have access to a calculator when dealing with numbers, so most numeracy tests allow you to use one too. In this way the test reflects the reality of real-life jobs.

The questions are designed to assess your understanding of the numerical information in the test. You are not required to make decisions or solve complex problems during the test. Your job is to demonstrate that you can interpret and understand the level of written information that you would encounter in the job itself.

Numeracy tests are generally used in recruitment and training processes for jobs that do not require high levels of numerical ability. The tests are used to determine whether you have the minimum level of numerical ability to deal with the aspects of the job that require you to work with numbers. For example, a numeracy comprehension test may be used in the recruitment process for retail assistants working in a supermarket. This job requires many capabilities (especially customer service skills) but only requires numerical ability at a basic level. The employer in this case would use the numeracy test to confirm that you would be able to perform tasks such as looking up prices, calculating the correct change to give to a customer and checking figures on a stock manifest.

The types of calculation you may be required to perform in a numeracy test include addition, subtraction, multiplication and division. You may be required to interpret numerical information presented in tables and graphs. You are unlikely to be required to perform more complex calculations such as percentages, averages or fractions.

See Chapter 3 for practice questions and advice on the best approach to taking a numeracy test.

INSIDER INFO

THE TRADER'S TEST

Usually when you take a test it will be in a quiet location that is free from distractions – this helps you to perform at your maximum level. But sometimes we develop tests that place greater demands on your numerical reasoning abilities. A number of years ago we developed a numerical test for graduate candidates applying for positions as financial traders on a busy trading floor. The questions in the test were pitched at a very high level to reflect the demands of the job and the time limit for the test was kept very short to represent the limited amount of time within which real traders must make complex financial decisions.

To make the test even more demanding, the testing session was conducted with loud dealing-room noise being played in the background. This simulated the demands of the real environment – trading floors can be very noisy.

The test therefore assessed the candidate's ability to perform very high-level numerical critical reasoning tasks under time pressure in a stressful environment. It also gave candidates a preview of the likely demands of the job.

Numerical reasoning tests

Numerical reasoning tests are the most widely used form of numerical ability test because they reflect the level of demands in the majority of jobs. The numerical information in these tests is likely to be more complex and harder to interpret than the content of a numeracy test. You will be required to perform more calculations to answer the questions and these will be more complex than those in a numeracy test.

A numerical reasoning test measures a more advanced type of numerical ability. Not only are you required to interpret and understand the numerical information, you also need to reason with it. This means you must make decisions based on what you understand the information to mean. To arrive at the correct answer you will often need to perform calculations that require a number of steps and that are more complex, such as percentages and averages.

A typical numerical reasoning question will require you to identify the correct pieces of information in the numerical data you have been presented with (interpretation) and then correctly work with these figures to arrive at the correct answer (calculation). Numerical reasoning tests obviously require a degree of numeracy that is more advanced than entry-level numeracy tests and the ability to perform higher level forms of calculation. You will also need to demonstrate a degree of attention to detail and focus to avoid making simple errors in your calculations.

Numerical reasoning tests are generally used in recruitment and training processes for jobs that require mid-levels of numerical ability. The tests are used with a broad range of educational levels, from school leavers to graduate level candidates. The broad nature of the ability measured by numerical reasoning tests mean that the questions within the test vary a great deal in difficulty; this means there are likely to be some questions you find harder than others to answer.

For example, some questions may require simple addition while others will involve multiple stages of calculations and techniques such as percentages, averages, ratios and fractions.

Numerical reasoning tests are used to assess your potential to deal effectively with a wide range of job demands. For example, a numerical reasoning test may be used in the recruitment process for call-centre staff. This job not only requires important attributes such as customer service skills but also requires you to demonstrate a level of numerical ability that is sufficient to deal with the more complex aspects of the role. The employer in this example would use the numerical reasoning test to confirm that you would be able to learn to interpret different customer pricing tariffs, calculate discounts, interpret account information as well as solve problems effectively and understand complex customer service issues.

See Chapter 4 for practice questions and advice for taking a numerical reasoning test.

Numerical critical reasoning tests

Numerical critical reasoning tests are the most advanced form of numerical test. These tests contain complex and high-level numerical information that is designed to simulate the demands of senior level jobs.

The questions are designed to assess your ability to draw conclusions and make complex decisions about the numerical information in the test. These decisions sometimes require you to use inference or deal with ambiguity in the data and base answers on a degree of estimation. Sometimes it is the sheer complexity of the numerical data that requires high levels of numerical ability to answer the questions correctly. Numerical critical reasoning tests can simulate the demands of working with complex balance sheets and profit and loss accounts.

INSIDER INFO

BESPOKE AND SPECIALIST NUMERICAL TESTS

Numerical tests tend to be fairly generic across different types of jobs or sectors, so the numerical tests used for any role, including more technical jobs in finance or IT, will generally be identical. Employers buy off-the-shelf tests from test publishers and this is why you will sometimes be asked to take the same test by different organisations.

Some employers will ask business psychologists to develop bespoke tests for the organisation. I have developed bespoke numerical tests for a range of larger organisations over the years based on numerical material that is specific to each job or organisation.

When developing bespoke or specialist tests we ensure that you do not need prior knowledge of the specific role or business to do well. We also make sure that the numerical content of the test is representative of the real demands in the job, which gives you, as the test taker, a valuable insight into whether or not it's the right job for you.

For jobs such as IT or finance, employers will often use numerical tests alongside other types of test, such as verbal critical reasoning or spatial reasoning.

The difficulty level of numerical critical reasoning tests is ratcheted up by requiring you to interpret more complex data as well as perform multiple-stage calculations in a systematic way in order to identify the correct answer. You are likely to make much more use of your rough paper when completing a numerical critical reasoning test. This is because you will need to write down many sub-totals and other results during the process of answering each question. It is good test-taking practice to write these values down since they often become useful later on in the test – it will save you time as you will not need to calculate them again.

INSIDER INFO

A STITCH IN 9 SAVES TIME

For higher-level numerical tests you will need to complete a number of calculations to answer each question. If the numerical information in the test is based around prices for example, many of the questions may require you to add all the prices together – to calculate an average or a percentage for example.

You will save yourself valuable time during the test if you write down any sub-totals that you calculate while completing the questions on the rough paper provided. This means that in the event of another question requiring the same sub-total (such as the sum of all the prices) as part of the calculation, you will already have it.

A couple of bits of advice though: make sure you label the sub-total so that you don't mix it up with any others. And make sure you've calculated it correctly – otherwise you'll make a mistake every time you use it.

Numerical critical reasoning tests are used in recruitment and training processes for jobs that require the highest levels of numerical ability. The tests are used to determine whether you have the mental firepower to deal with the complex nature of the job. For example, a numerical critical reasoning test may be used in the recruitment process for graduate trainee accountants, a job that requires numerical abilities of sufficient power to understand complex figures and data, solve difficult problems and perform demanding analyses.

See Chapter 5 for practice questions and advice on passing numerical critical reasoning tests.

Numerical test materials

A valuable piece of preparation for a numerical test is ensuring that you are familiar with the format of the materials you will use to complete the test. Most modern numerical tests follow a similar approach to the format of the paper and pencil test materials. When you complete the test you will be given everything you need, which should include:

▶ Test booklet – this contains the instructions for taking the test and the test pages themselves with written information and questions. You should not write on this booklet as it is designed to be reusable

▶ Answer sheet – this is what you use to record your answers

▶ Calculator – this is provided by the test administrator; you do not need to take your own since you will not be allowed to use it

▶ Pencil and eraser or a pen

▶ Rough paper – to use for working out calculations.

People who have not taken many tests in the past sometimes find the plethora of test materials confusing. If you can prepare yourself by familiarising yourself with knowledge of the format of the test materials you will be less likely to become confused or flustered in the heat of the testing session.

Most paper and pencil numerical test materials follow the format described here. You can take your preparation further by finding out more about the specific numerical test you are due to complete.

You may be asked to complete an online numerical test. Online tests follow a different format from traditional paper and pencil tests – except you get to use your own calculator!

See 'Find out more about your specific test' later in this chapter for more. See Chapter 6 to find out more about online testing.

Some myths about numerical tests

Even those of us who left school many years ago may still have fond memories of the numerical tests we were asked to

INSIDER INFO

OLD-SCHOOL NUMERICAL TESTS

Modern occupational tests have their roots in older educational tests. In fact very early occupational tests of ability look a lot like their educational forebears. If you are asked to complete an ability test that looks and feels like an old school test you may form the impression that the potential employer is either a bit old fashioned or they have simply not updated their recruitment processes for many years. The test itself may be outdated and the results somewhat questionable.

Modern professional employers tend to use contemporary numerical tests that reflect the style and demands of their workplace. If you are asked to complete an old-fashioned looking test it might be because the employer themselves are somewhat of the old school – this might influence your decision about whether or not the job is right for you.

complete during our education. Sometimes these were maths exercises or perhaps simply being asked to answer a numerical question from an example that your teacher was talking through with the whole class. For most people educational testing is the first experience of being tested, and often it is the experience of being put on the spot or trying to do sums under pressure that lies beneath the anxiety many people feel about numerical tests.

To practise and pass a numerical test for an employer, you need to let go of any preconceptions that you have about testing based on your experiences at school. Modern tests that have been designed to be used in a workplace have little in common with those used in schools. There are a number of myths about testing that are based on educational experiences:

Numerical tests you'll never see in real life

Modern employers do not use numerical tests that are based on school-like formats such as:

- ▶ Adding together lots of fractions
- ▶ Solving numerical problems that use lots of words (e.g. numerical problem-based questions such as: It takes 3 men 2 hours to dig a hole 2m deep; how long would it take 2 men to dig a hole 4m deep?)
- ▶ Long division by hand
- ▶ Completing sequences of numbers
- ▶ Times tables
- ▶ Geometry – calculating areas, angles, volumes
- ▶ Puzzles based on diagrams and shapes with numbers in them.

These formats have fallen out of favour with employers because they rely too heavily on learned information – such as your command of the formula for using pi or method for calculating long division. These older forms of numerical testing also tend to over-rely on verbal content; the questions themselves contain a lot of words and language that can be tricky to interpret. This non-numerical content contaminates the testing

INSIDER INFO

INAPPROPRIATE TESTING

There are employment law implications for using tests that contain inappropriate content. In order to be a fair assessment of your suitability to do the job the test must contain tasks and demands that reflect the content of the job.

Employers tend to avoid tests that contain school-like content because they do not match the content of the job. Modern numerical tests reflect occupational life very closely.

Employers use tests that contain content such as long devision or non work-like calculations at their peril. If they are ever challenged over the fairness of a recruitment decision their use of such tests would be hard to defend.

of your numerical ability because you are forced to draw on other abilities, such as your verbal ability or spatial ability if the test uses lots of shapes. Numerical tests that adopt older testing formats do not measure your raw numerical ability in a way that is fair to all candidates from all backgrounds.

Numerical tests that you are asked to complete for an employer should be based on everyday, business-related numerical information, which you use to correctly answer questions using a multiple-choice format.

Revision for tests

You cannot revise for a numerical ability test. Modern tests require no prior knowledge and so you can't revise for them in the same way that you do for a school exam. You can prepare for numerical tests – which is why you're reading this book!

No long-hand written answers

Modern numerical tests do not require you to write long answers like school exams. Numerical tests use a multiple-choice format to make scoring easy and fair. You will not be asked to show your working (but any rough paper you use to help with your calculations will be collected at the end of the test. This is to make sure you do not try to smuggle your answers out of the test room; your working is not scored).

No memory testing

You are not required to memorise the numbers and figures used in a numerical test. You can refer to the information in the test booklet to answer the questions throughout the test session. Some newer numerical tests even give you reading time before the test begins to allow you to familiarise yourself with the numerical information.

No negative marking for wrong answers

Contemporary numerical ability tests do not cut marks for wrong answers. This is an approach sometimes used by old-fashioned educational tests to try to compensate for people guessing the correct answers. By taking points away for wrong answers educational testers believed they would deter guessing.

INSIDER INFO

BIAS IN TESTING

Modern tests have to perform against stringent standards of fairness. All candidates regardless of background should have an equal chance to perform well on the test. Early tests were often biased against certain groups. IQ tests became controversial because some groups of people scored higher than others – white middle-class men being the highest scoring group. This led a number of psychologists to incorrectly believe that there were real differences in IQ between groups – in fact it was simply the result of bias in the test.

Modern tests are developed to minimise such bias by ensuring that the cultural, educational and social background of candidates does not confer an advantage or disadvantage.

The scoring process for modern, occupational numerical tests is kept secret by the test developers to prevent cheating and there may be a correction made to test scores to compensate for possible guessing, but modern tests do not subtract scores for answering incorrectly.

Find out more about your specific test

Your knowledge of numerical tests will ensure that you are better prepared for the testing session. By letting go of any testing myths and understanding the format and materials of numerical tests you will be in a much better position to benefit from your efforts to practise and develop your test-taking strategies.

You can develop your knowledge further by finding out more about the specific test you will be taking. This will ensure that you are prepared to deal with anything unusual or idiosyncratic about the precise test you are asked to complete. Below are a number of sources of information for specific tests.

▸ **Ask the employer**. The organisation that has asked you to complete the test should provide you with details about the specific assessment it uses. It is part of the testing best-practice code for employers to ensure

INSIDER INFO

SPEED TESTS VERSUS POWER TESTS

Tests with a generous time limit are called 'power tests'. They are designed to measure the power of your numerical ability – in other words they establish the maximum level of difficulty you can perform at. A power test is designed so that most people have enough time to answer all the questions before the end of the test, but the questions get progressively more difficult as you complete the test – the final questions are much harder than the early ones.

Speed tests set a very tight time limit – they are not designed to help you to complete all the questions before the test ends. These tests measure the speed of your numerical ability – how quickly you can answer questions of equal difficulty. The questions do not become harder as you progress.

A typical numerical ability speed test might have 30 questions but only a 15-minute time limit, while a numerical ability power test might have 30 questions and a 30-minute time limit. You can determine whether a test is more speed or more power by looking at this ratio of number of questions to amount of time allowed. If the test allows less than 30 seconds per question it is more likely to be a speed test. This would be 60 seconds for a critical reasoning test, where the high-level questions take time to calculate.

Advice for how to adapt your approach to taking speed or power tests can be found in the test-taking strategies section of this chapter.

that their test candidates are fully informed about the testing process.

▶ **Test practice materials**. Test publishers offer practice materials for the tests they provide to employers. These practice materials will build on the information and advice in this book by providing guidance and example questions specific to the test you will take. Although it is not compulsory for employers to send practice materials to their test candidates, it is best practice. If a prospective employer does not provide any practice materials prior to testing you could contact them to find out who publishes the test they will be using. You can then contact this test publisher for advice about the test. Many test publishers provide free practice materials specific to their tests. See Chapter 9 for a list of test publishers that offer practice questions on their websites.

▶ **British Psychological Society** (BPS). Visit the BPS website (www.psychtesting.org.uk) for advice to candidates taking a test.

See also Chapter 9 for more resources.

STEP 2: STRATEGIES – LEVELLING THE PLAYING FIELD

The second step to achieving your personal best score on a numerical test is to develop your test-taking strategies. Your approach during the test session can make a big difference to your test score. Candidates who have poor test-taking strategies are much less likely to achieve their personal best score – regardless of their knowledge of the test and the practice they have done. If you are unprepared in terms of your test-taking strategies then you are unlikely to demonstrate your maximum level of ability.

How test strategies can help you

Experienced test takers tend to have high levels of test sophistication – they know the best approach to take during the test session. Taking lots of tests gives them an advantage over less experienced candidates because of the strategies and techniques they have developed though their exposure to tests.

You can level the playing field by developing these test-taking strategies yourself. This will enable you to compete with the most experienced test takers and maximise your own score. Research into test performance has shown that increased test sophistication can lead to significant increases in test scores. The more valuable test-taking strategies demonstrated by test-sophisticated candidates are described below.

Before the test session

Make sure that you are in peak form for completing the test. Get a good night's sleep before the session. Arrive at the testing location in plenty of time – you don't want the stress of a delay distracting you from performing at your best. If you need spectacles for reading you will need them for the test.

Listen very carefully

The test administrator will read aloud from a test administration card; you will be asked to read the instructions from your test booklet at the same time. Make sure you focus on what the administrator tells you – do not let your mind wander to try to read ahead to the later stages of the instructions or you may miss something important.

Ask questions

If there is anything that you do not understand, make sure you ask about it before you start the test. Do not feel nervous about asking questions – you can avoid making silly mistakes by making sure you completely understand what you need to do during the test. In most test sessions you are not allowed to ask questions once the timed part of the test has begun; so make sure you clear up any uncertainties you have in the beginning part of the test session.

Pace yourself

Make sure you know how much time you have for the test so that you can pace yourself. Most sessions do not have any reminders or warnings about the amount of time you have left. You can write a note of the time that you started the test on your answer sheet or any rough paper you've been given so that you do not forget. For tests that look like power tests (generous time limit for the number of questions) you should not set yourself a time limit for each question because some will be harder than others. For speed tests (short time limit for the total number of questions) you need to work quickly but accurately – but do not expect to finish them all.

Concentrate

Work as quickly and accurately as you can and do not get distracted. Do not be worried by what other candidates are doing. Focus on your own performance. If you are remotely completing the test online, try to use a quiet location free from distractions.

Do not get stuck

If you find a particular question difficult do not spend too much time on it. Leave it and go on to the next. If you have time at the end you can always come back to it. You should try not to leave any questions unanswered – make a note of any questions that you have not completed so that you can return to them at the end of the test (before the time limit for the test is reached).

Avoid guessing

When you are not sure of an answer do not simply guess randomly. Try to narrow down the options by deciding which of the answers, true, false or cannot say, are most likely to be incorrect.

From the two remaining options try to identify which is the most likely to be correct. It is better to use this 'best guess' than leave a question unanswered.

Revisit your answers

If you complete all the questions before the end of the time allowed, use the time you have left to revisit and check your answers. This is especially important for numerical tests where a simple error of arithmetic can cause you to get the answer wrong. It is worth checking all your answers if you have time, but you should focus particularly on questions where you struggled or had to give your best guess.

Watch out for sausage fingers

Many errors made by candidates when completing numerical tests are simply the result of miss-keying numbers into the calculator. It seems an obvious point, but you should take care not to rush your use of the calculator or press buttons by mistake (as if you had fingers like sausages).

Don't be easily distracted

The wrong answers in multiple-choice formats are called 'distracters' (see page 23 for more information); they are there to tempt you towards selecting a wrong answer if you have made an error in your calculations or reasoning. Rather than simply selecting the first answer that looks correct you should check all of the choices to make sure that you have not fallen into the trap of picking the wrong answer. An example is given below.

Drink	Price
Café latte	£1.90
Cappuccino	£2.10
Hot chocolate	£2.00
Tea	£1.90

INSIDER INFO

SAUSAGE FINGERS

Not the same as 'butter fingers,' but close when it comes to taking a numerical test. The term 'sausage fingers' relates to careless use of a calculator or keyboard, where your finger hits the wrong key or maybe two keys simultaneously – as if you've got really huge sausage fingers.

When taking tests, a few episodes of sausage fingers can cost you precious time as you are forced to re-key your calculations to arrive at the correct answer. The problem can be exacerbated by nerves, which adversely affect your finger dexterity and cause you to rush – two prime causes of sausage-fingered miss-keying effects.

Sausage fingers are of course not solely the preserve of sweaty-fingered test takers. Banks, especially those with trading and brokerage divisions, often insure themselves against any miss-keying of data, which could result in rogue trades potentially leading to multi-million pound losses.

When using the calculator (or keyboard if it's an online test) during a test remember to slow yourself down and keep your eye on the ball – your dexterity and error avoidance rates will both increase if you watch what your fingers are doing.

If you really do have big fingers why not use the end of your pencil?

Cross culturally 'sausage fingers' takes on a different meaning; in Korea, 2009 sales of snack sausages increased by almost 40% in the winter as iPhone users utilised the sausage as a stylus in order to avoid removing their gloves in the cold. That particular brand of sausage proved to be compatible with the phone's electrostatic touch-screen.

Please note that you're unlikely to be permitted to take Korean snack sausages into the testing room, regardless of how much more effective they may prove to be than your own fingers.

1	A customer buys 2 cups of the same drink and spends £3.80. Which type of drink does the customer buy?			
A	B	C	D	E
Café latte	Cappuccino	Hot chocolate	Tea	Cannot say

Candidates with poor test-taking strategies may fall into the trap of picking answer A. This is because they quickly work out the price of a single drink (£3.80 ÷ 2 = £1.90) and see that this is the price of a café latte, which is the first of the multiple-choice distracters, and tick answer A. A candidate who takes a moment to check all of the choices is more likely to get the correct answer. The thought process that a candidate with good test-taking strategies will follow is as follows.

▸ **Step 1**: Calculate price of a single drink bought by the customer (£1.90).
▸ **Step 2**: Check distracter A – café latte is £1.90 so this is a possible correct answer.
▸ **Step 3**: Check distracter B – cappuccino is £2.10 so B is an incorrect answer.
▸ **Step 4**: Check distracter C – hot chocolate is £2.00 so C is an incorrect answer.
▸ **Step 5**: Check distracter D – tea is £1.90 so D is also a possible correct answer.
▸ **Step 5**: Check distracter E – the customer could have bought 2 cups of either A or D, therefore E – cannot say – is the correct answer.

You can see that this candidate avoided being distracted by the first incorrect multiple-choice answer by adopting a systematic approach to checking each choice before answering.

This technique also works well when you find a question tricky. You should check each of the multiple-choice distracters in turn, ruling out any answers that are clearly wrong (such as answers B and C in the example above). Then you can focus on identifying which one is correct. Often only two of the multiple choices will look like possible correct answers or options – so at least you have a 50:50 chance of identifying the right one!

Speed versus power

Some tests are speeded – they have a short time limit and a lot of questions; candidates are not expected to attempt all the questions within a speed test. Other numerical tests are power tests – they have a more generous time limit for the total number of questions and are designed to enable the vast percentage of candidates to attempt all the questions within the time available.

You can identify whether a numerical test is speed or power test by applying a rule of thumb to the number of questions in the test compared with the time limit. If you are allowed less than 30–60 seconds per question then it is more likely to be a speed test. You can also sometimes tell by looking at the questions at the end of the test. Power tests tend to increase in difficulty as the questions progress. The final question may look a lot harder than the questions at the beginning. It is much easier to spot an increase in difficulty with a numerical test than it is with some other types of test, such as a verbal test. Here's a real-life example from a graduate level numerical critical reasoning test:

First question: 'What is the average cover price of the 8 magazines?'

Final question: 'The music magazine Jazz has one half the number of male readers that Utopia has, but for Jazz the ratio of male to female readers is 3:2. Estimate the number of female readers of Jazz.'

You can see that the final question in the test is substantially more demanding than the first. This particular test has

30 questions and a time limit of 45 minutes. It is almost certainly a power test. (In fact with this test 98.7% of candidates attempt all the questions within the time limit.)

You should adjust your test-taking strategy depending on whether the test looks like a speed or power test. With power tests the questions are likely to become more difficult as the test progresses, but you should have time to attempt all the questions in the test. This means that you should adjust your test-taking strategy for a power test in the following ways:

▶ **Pace yourself**. You do not need to rush but you shouldn't dwell too long on any question where you become stuck. Leave the question until you have attempted all the other questions in the test – you are likely to have sufficient time at the end to return to the question and try again. Many people find that leaving a question helps clear their thinking, which makes the answer more obvious when you try again. Perhaps you unconsciously carry on working on the question in the background while your rational mind focuses on the other questions?

▶ **Expect to slow down**. As the questions become harder your pace will decrease. This is normal so you should allow yourself greater time for the more difficult questions as you get towards the end of the test. Do not try to maintain the same pace throughout the test.

▶ **Use any extra time well**. The more generous time limits of a power test mean that you may have some time left over at the end once you've attempted all the questions in the test. Use this time well to check your answers and revisit any questions you struggled with initially.

Speed tests require a different focus for your test-taking strategies. With a speeded numerical test you probably won't have time to attempt all the questions in the test and may run out of time before you finish. The questions themselves are less likely to increase in difficulty in the same way that the questions do in a power test. These features of a speed test mean that you need to adjust your test-taking strategy in the following ways.

▶ **Pace yourself**. Your score on a numerical speed test will be heavily influenced by the number of questions you manage to answer in the time available. But you also need to answer the questions correctly – there are no marks for wrong answers no matter how many you answer within the time limit. This means that you have to pace yourself effectively. The most important mistake to avoid is spending too long on a question that you get stuck with. Do not waste precious minutes struggling with a question; you could be using this time to answer other questions correctly and pushing up your test score. If you do not know the answer to a question you should apply the usual test-taking strategy of giving the answer you think is best without simply guessing (see page 44).

▶ **Don't panic**. Some people feel under additional pressure when taking a speed test because of the sense of the insufficient time period ticking away in the background. This is bad news as panic can affect your concentration and ability to answer questions correctly. Remind yourself that the time limit is part of the test and therefore affects everybody who takes the test. You are not expected to attempt all the questions in the test and you should not try to. Instead you should focus on calming yourself and concentrating on the task at hand, which is answering each test question in turn. For advice on managing your emotions in the test session, see the next section.

▶ **Don't cheat!** Some people are tempted to use the closing seconds of the time limit to quickly fill in the answers to questions that they have not yet attempted.

The temptation is understandable; if there are five choices then you would have a 1 in 5 chance of guessing each answer correctly. Be aware though that the scoring mechanism for tests can include a correction for this type of test-taking behaviour, which would adversely affect your test result. The cost of randomly guessing some questions at the end of the time limit could be a worse test result than if you'd simply let things be and had your score based purely on the questions you attempted legitimately.

Whether a test is a speed one or a power one your main focus should still be on answering the questions correctly –this after all is what drives your numerical test score. Do not worry and fret after the test if you did not answer all the questions or you are unsure of some of your answers. The scoring and time limits of numerical tests are carefully designed to ensure an accurate measurement of your numerical ability. Provided you applied the appropriate test-taking strategies your performance should reflect your personal best score.

Calming and focusing strategies

In addition to developing your test sophistication through the strategies you adopt during the test session it is important to make sure that your mental state is focused on achieving your personal best score. Prior to the test session you should spend some time preparing your mind and emotions. To perform your best during the test session you need to feel calm, focused and at your mental peak.

Many people feel anxious before and during a test session, especially when there is a lot riding on the results. Numerical tests in particular seem to be a source for a lot of dread for many people. You need to deal with these nerves to prevent them from affecting your performance. You can draw from the techniques described below, which are often used by sports psychologists when helping athletes perform at their peak performance:

Centre your breathing

Prior to and during the test session you can centre yourself and minimise any nerves by focusing your attention entirely on your breathing. Still your mind and quieten worries by simply extending your breaths and using the full extent of your lungs. Try to make each breath softer and longer than the last.

Look at the trees

Before attempting a challenge sportspeople will often 'look at the trees'; they take a moment to look into the distance and take a moment away from the immediate situation and its demands. You do not need to be near any trees or even outside – simply take a moment to gaze into the middle distance before embarking on the challenge.

Focus on your objectives

Centring your breathing and 'looking at the trees' give you a valuable opportunity to focus on your objectives. In that moment of calm you should remind yourself what you are here to do. Do not allow your mind to worry about the outcome; instead rehearse in your mind what you are going to do during the test session.

Remember a calm place

Remove your mind from the stress of the situation by bringing to mind a place where you have felt calm. This does not have to be a real place but you need to be able to imagine it in as much detail as possible. Distract your mind from the worries taking up space by running through each of your senses – remember how the calm place looks, feels, smells and sounds.

Name your feelings

Focus your attention on the physical sensations of your body. Note any feelings that the anxiety is causing. Explore each part of your body and note any tension, discomfort, heat or any other sensations. Each time you discover a symptom you should spend some time breathing it away.

Get some help

If your test nerves are extreme then you could consider seeking some more professional help. Many people have found hypnotherapy a useful aid to controlling nerves and anxiety in pressured situations like tests, exams and interviews. A qualified hypnotherapist will be able to teach you techniques to get your feelings under control and focus your energies on maximising your performance.

Don't panic!

This is the best advice of all. Do not allow yourself to be sucked into a spiral of anxiety by a difficult question. Some candidates get stuck on a question and spend far too much time trying to work it out. Often their thinking has been affected by anxiety – rather than thinking through their answer they are worrying about getting it wrong. This worry spirals into anxiety where the candidate begins to catastrophise their thinking into believing that they are going to fail the whole test. A fog descends over the candidate's mind and they begin randomly tapping numbers into the calculator hoping that one of the multiple-choice answers will magically appear.

If this sounds like an experience you've had in the past with a numerical test then remember to avoid getting stuck in this trap in future. When you're stuck it is always best to move on. If there's time at the end of the test you can try it again. Often, like the lid on a jam jar, the answer is easier to unlock when you tackle the question afresh.

INSIDER INFO

FEAR AND LOATHING IN MATHS TESTING

Anyone who uses psychometric tests regularly will tell you that numerical tests are the single biggest source of anxiety and nerves among test takers – far more than tests of other abilities such as verbal, mechanical, spatial or abstract reasoning. Yet scores on numerical tests are no lower across the general population than on other types of test – just as many people do well, or badly, on numerical tests as on other types of test.

This phenomenon is sometimes called 'maths anxiety'; it is thought to be caused by a range of factors specific to common maths learning experiences:

▶ Right or wrong answers – and the fear of getting it wrong
▶ Beliefs about maths abilities – that only 'gifted' people can truly do well at maths (nerds!)
▶ Fear of public embarrassment – not being able to come up with the correct answer in front of others
▶ Negative attitudes to maths – that is, it is irrelevant to 'real life'
▶ Teachers infecting students with anxiety – especially in complex areas of maths that are tricky to explain
▶ Rote learning and practice approach to teaching maths – does not work for all learning styles.

Whatever the source of maths anxiety, it can become a problem if it continues to affect your performance on numerical tests. People who experience anxiety and nerves with maths exams will often transfer the same emotions and physical feelings into the numerical test session, where it is likely to impair their performance.

There is advice on how to tackle testing nerves in this book. But with numerical testing nerves, there is one more piece of advice that is often useful in tackling deep-seated maths anxiety. A valuable technique for controlling your numerical testing anxiety is to rationalise it; ask yourself where it comes from. You might be able to identify earlier situations in life that are related to your experiences of maths. You should aim to externalise your anxiety as something that used to affect you but was never part of you. Explore ways in which your life has changed since those earlier situations and as a result how the anxiety is no longer relevant.

If you don't get time to come back then all you have done is lost one point in the test – this is less likely to have a big effect on your overall test result than spending too long on a single question and running out of time.

Test session survival strategies

Tests are generally administered to small groups of candidates at the employer's offices. Online tests are usually administered remotely – in other words you complete them at home or at your office in your own time.

See Chapter 6 for more information about online testing.

A traditional, supervised testing session itself can be very nerve-racking – not least because for many people it is an unfamiliar and confusing process. However, most testing sessions follow a similar structure that is designed to standardise the way in which tests are administered. This ensures testing is fair by introducing consistency across all test sessions – everyone who takes the test will experience the same conditions. By understanding this structure you can combat many of the nerves associated with testing. You can also ensure that your results benefit from the elements of the testing structure that are designed to help you.

The test session structure

The testing session will be delivered by a trained test administrator who will follow a carefully defined set of steps for the specific test they are using. If the numerical test is being used alongside another ability test the administrator will often sequence the session so that the numerical test comes last. This is because numerical tests seem to create more anxiety in many candidates than other types of test; so if something like a verbal test is also being used it will be administered first to allow candidates to 'warm up' before the numerical test.

INSIDER INFO

THE TEST ADMINISTRATOR

Test administration is a skilled job and requires specialist training. Anybody involved in the use of ability tests should be properly trained to administer and score the test. You can check the testing qualifications held by an employer by contacting the British Psychological Society's Psychological Testing Centre (see Chapter 9 for details).

The test administrator's responsibilities are to ensure that the test session is as standardised as possible – in other words each session is identical to the last. The test administrator must also ensure that every candidate is able to perform at their best and is not affected by nerves or a lack of understanding about the test.

▶ **Introduction** – the test administrator will set the scene and tell you more about the test session. You should use this part of the process to ask any questions you have about the test. Remember that asking questions when uncertain is a key test-taking strategy.

▶ **Instructions** – the test administrator will read the test instructions aloud from a card. These instructions will be reproduced in your test booklet and you should read along when asked to do so. Clear your mind and read along carefully to ensure you do not miss any details that are important. Avoid reading ahead or allowing your mind to wander. Remember that paying attention and focusing are key test-taking strategies. You can still ask questions at this stage of the process.

▶ **Example questions** – at the end of the instructions the test administrator will ask you to complete some example questions. These will not contribute to your test score and they will not be timed. They are designed to help you understand the format of the test. You can

INSIDER INFO

TEST SECURITY

Test administrators will collect up all the materials before you are allowed to leave the testing room. This is to maintain the security of the test materials and to prevent them from being circulated among future candidates. Even rough paper is collected – candidates have been known to write the test questions and answers down to share with their friends.

ask questions during this part of the process. If you have any difficulty with an example question the test administrator can help you to find the correct answer. Do not be worried if you answer an example question incorrectly – many people do because the questions are often designed to illustrate any tricky or complex aspect of the testing format (such as how a cannot-say option works).

▶ **Before you begin** – after the example questions, the test administrator will read the final part of the test instructions from the card. This will include advice about how to do well during the test as well as remind you about the time limit. This is your last chance to ask any questions.

▶ **Take the test** – the test administrator will tell you to begin the test and will start the timer. You will not be allowed to ask questions once the test has begun. The administrator may not alert you before the time is up so remember to make a note of the time you started.

▶ **Time's up** – the test administrator will ask you to stop when the time limit has been reached. The test administrator will collect up all the test materials before thanking you and inviting you to move on to your next activity.

THE TESTING LOG

Test administrators will record any unusual behaviour or events that they observe during the test session. Sometimes this might be information relating to the conditions of the test session – if a car alarm was going off during the test for example. They will also record the candidates' behaviours, such as attempts to complete questions after the time limit. Any information recorded by the test administrator may be used to guide the interpretation of the test results for an individual candidate or the whole group.

STEP 3: PRACTICE – DOESN'T MAKE PERFECT (ON ITS OWN)

The third and final step to achieving your personal best score on a numerical test is to practise. Remember that practice alone is not enough to perfect your test score – you also need to devote time to building your knowledge of the test and developing your test-taking strategies.

With knowledge and strategies in place your practice will reap you the maximum reward in terms of your test score. This combined approach is critical to achieving your personal best score.

What kind of practice helps?

The best type of practice is to complete example questions that are very similar to those in the specific test you will be taking. The employer who asks you to complete a test may provide you with some example questions developed by the test publisher who built the specific numerical test you will be completing.

INSIDER INFO

TAKING A TEST MORE THAN ONCE

Research conducted in the early 1990s identified the existence of 'practice effects' – the factors that affect people's test scores when they take the same test more than once. This can sometimes be a problem for test developers and employers when they are testing candidates who are applying to lots of employers who all use the same test. Employers worry that people's test scores will increase because they remember the test and its questions from previous attempts. Some test publishers are keen to downplay the likelihood of a change in test scores following repeat administrations.

The research, which focused on graduates – who are the most frequently tested and re-tested group, found that people's test scores tend to change when they take the same test twice. Many people's scores increase, but some stay the same and some decrease. The research ruled out the influence of memory and instead identified motivation, confidence and test sophistication as the key factors behind practice effects.

While practice effects happen with all types of test, this study found that numerical test results saw some of the biggest increases due to practice effects. If you are asked to complete a numerical test you can make big improvements to your personal best score by working through this book! This can help you develop confidence and test sophistication. To avoid a decrease in your test score you must maintain your motivation – do not become complacent or careless when taking a test no matter how confident you feel.

The practice questions in this book are another valuable resource. Unlike many of the other numerical reasoning test books, the questions in this book follow the same format as the questions in modern numerical tests. Whether you use these in conjunction with example questions provided by the employer or as an alternative, these practice questions will enhance your ability to perform well on the test.

Practice questions help you improve your score in three ways.

▸ Completing the questions increases your knowledge of the test format.
▸ Advice on how to answer the practice questions correctly helps you refine your test-taking strategies.
▸ Completing the questions can help you develop the underlying ability measured by the test – so answering numerical test questions can help you develop your numerical ability.

The practice questions in this book cover the range of levels of difficulty of different numerical tests. Working through these questions will help you develop your personal best test score.

Developing your raw numerical ability

There are additional forms of practice that can help you to develop your personal best score. Your numerical ability is like a muscle – and like any muscle it responds to exercise. Regularly exercising your numerical ability will help it to grow and become stronger, and if you do not, it will become weaker.

Completing practice questions will help you to exercise and therefore develop your numerical ability but the benefits are limited. To build your raw ability to its maximum strength you need to combine your practice with regular use of the ability. Practice combined with regular exercise will encourage your numerical ability to become stronger and well-rounded.

Your raw numerical ability can be exercised and developed through any activity that requires you to use it, such as:

▸ Completing number puzzles (such as Sudoku), word games, problem-solving challenges
▸ Adding things up in your head – such as prices as you go around a supermarket

- Reconciling your current account at the end of each month
- Interpretation – review complex numerical data, such as a set of accounts or financial report at work
- Analysis – interpret some numerical information and then spend some time analysing it. Identify key themes, changes (such as from year to year if it is annual accounts) or look for where the biggest differences occur between values (such as costs and sales)
- Taking part in discussions or debates about numerical information

Like any form of exercise these activities will be most successful when you integrate them into your normal life. If you can perform the exercises in a way that becomes a habitual part of your day, your raw numerical ability will grow steadily. It is much better to introduce small changes, such as taking the time to complete a daily Sudoku puzzle, than attempt infrequent or overly demanding numerical exercises.

Remember that there is an upper limit to your numerical ability. Just as we can't all grow our muscles to the same size as Mr Universe we can't all grow our numerical ability to the level of Stephen Hawking. What you can do is combine knowledge of the test, your test-taking strategies and practice to develop your numerical ability to its maximum – this will help you achieve your personal best score.

CHAPTER 3
NUMERACY

Numeracy tests are the simplest form of numerical ability test. These tests are based on clear and straightforward numerical information and a series of questions requiring interpretation and simple arithmetic.

The questions are designed to assess your understanding of the numerical information. You are not required to make decisions or solve complex problems during the test. Your job is to demonstrate that you can interpret and understand the level of written information that you would encounter in the job itself.

GETTING THE MOST FROM THESE PRACTICE QUESTIONS

These numeracy practice questions are grouped together into three short practice tests. Each practice test mirrors the approach of a real test; there is a short extract of numerical information followed by a series of five questions for you to answer about the information you have read.

You're encouraged to complete each practice test one at a time and to spend time at the end of each short test to review your responses against the correct answers. Once you are happy with your work, move on to the next test.

The test instructions for these practice tests follow the same format and approach of instructions in real tests – to make sure that you are prepared for the actual tests, make sure that you familiarise yourself with these.

Completing the practice tests

Ideally you should complete the practice tests in conditions that are as close to the real testing environment and experience as possible.

▸ Find a quiet place that is free from any distractions.

▸ Read the practice test instructions before beginning the first practice test.

▸ Real tests have strict time limits. Simulate this by giving yourself five minutes per practice test. Start timing from the point you begin reading the numerical information for the test.

▸ You may use a calculator and rough paper to help with your working (do not use a complex, scientific calculator – only basic pocket calculators are allowed in real tests).

▸ Mark your answers in the spaces provided beneath each practice question by circling the answer you think is correct, A, B, C, D or E.

▸ Do not look at the correct answers until you have completed all the practice questions.

Advice for numeracy tests

Remember that while numeracy tests are the simplest type of numerical ability test, you should not assume they are straightforward. You need to approach these tests carefully, as you would any other, since they can place significant demands on your powers of attention to detail and ability to correctly interpret numerically presented information.

Numeracy tests require you to use the fundamental elements of numerical ability; the capability to accurately interpret an extract of numerical information and use it to answer questions correctly. These skills are critical to most jobs, not just those with minimal requirements for numerical ability.

Do not be tempted to skip these practice tests and to jump straight to the more difficult numerical reasoning and critical reasoning practice tests. Completing these numeracy practice tests will greatly benefit your test-taking strategies and level of practice.

When completing a numeracy test, remember these points to perform at your personal best.

▸ **Read the numerical information carefully** before beginning the practice questions. You do not need to memorise it but you will perform better if you are familiar with the information when attempting the questions.

▸ **Never simply write down your answer without checking it first**. Even if you think you have definitely calculated the correct answer you must take a moment to check the other choices to make sure you have not been distracted into selecting an incorrect answer.

▸ **Do not simply guess your answers**. You may think that you have a 1 in 5 chance of getting a question correct by guessing, but there is often a scoring mechanism designed to detect or penalise attempts to guess. If you cannot answer a question then review each of the multiple-choice options to identify the ones unlikely to be the correct answer. Then you can focus on the shortlist of possible correct answers to identify which one you feel is the right answer.

▸ **Use any time remaining to check your answers**. If you complete all five questions in a practice test within the five minutes time limit you should re-check your answers, as you would in a real test. This is good test-taking strategy, and you should aim to make it a habit.

Before you begin the practice tests

Read the test instructions carefully before you begin – these instructions have been adapted from real test instructions. So you will benefit from familiarity with the language and information they include at the time you actually need it during the test.

Reviewing your answers

Once you have completed a practice test you can turn over the page to check your responses against the correct answers. Along with each correct answer we have provided some tips and explanations about the answer.

Although these practice tests are too short to make an accurate measurement of your numerical reasoning ability – that's what real tests are for – we can provide some guidance about what your result means.

Score of 5	Score of 4	Score of 3	Score of 2	Score of 1
Great performance	Good performance	Average performance	Below average performance	Poor performance

Review your pattern of incorrect answers to see whether you are making some common mistakes. This might be simple errors of arithmetic, carelessness with the calculator or a tendency to fall for the distracters rather than selecting the correct answer. Understanding that you have a tendency towards a certain type of error should prompt you to be extra careful when answering questions that you find more difficult. For these questions you should pay close attention to what the question is asking, which piece of numerical information you must base your answers on, and how you are ruling in/out multiple-choice distracters.

Once you have reviewed your answers take a short break before attempting the next test.

NUMERACY
PRACTICE TESTS

PRACTICE TEST INSTRUCTIONS

These practice tests each consist of **five** questions
about the numerical information given on the first page of each test.
You are required to use the numerical information to answer each of the
questions in this test. For each question there are five choices of answer
labelled A–E – you need to choose one of them. Then mark your chosen
answer by circling in the appropriate option: A, B, C, D or E.

You may use a calculator and rough paper to help with your working-out.

▸ Allow yourself a 5-minute time-limit for each of the practice tests.
▸ You should work quickly and accurately.
▸ If you are not sure of an answer, fill in what you think, but do not
  simply guess your answers.
▸ Check your answers against the correct answers given on the
  pages that follow each practice test.

Remember these important points:

▶ Complete each practice test in a quiet place that is free from any distractions.

▶ Start timing the practice test from the point you begin reading the numerical information.

▶ Mark your answers for each practice question by fully circling the answer you think is correct.

▶ Remember you can use a pocket calculator and rough paper to help with your answers.

▶ Do not turn over to the correct answers until you have completed all the practice questions.

Now do the first practice test: start timing yourself and read through the numerical information first. Start answering the questions as soon as you are ready.

PRICE CHANGES

Drink	Old price	New price
Café latte	£1.90	£2.00
Cappuccino	£2.10	£2.20
Hot chocolate	£2.00	£2.20
Pot of tea	£1.90	£1.90
Bottled water	£1.50	£1.65

1 What is the new price of 1 cup of hot chocolate?

A	B	C	D	E
£1.65	£1.90	£2.00	£2.10	£2.20

2 What is the old price of 1 bottle of water?

A	B	C	D	E
£1.50	£1.65	£1.90	£2.00	£2.10

3 What is the price of 2 cappuccinos at the old price?

A	B	C	D	E
£2.10	£2.20	£3.80	£4.20	£4.40

4 Which drink has the biggest price increase?

A	B	C	D	E
Café latte	Cappuccino	Hot chocolate	Pot of tea	Bottled water

5 Which drink has the smallest difference between its old and new prices?

A	B	C	D	E
Café latte	Cappuccino	Hot chocolate	Pot of tea	Bottled water

PRICE CHANGES

Drink	Old price	New price
Café latte	£1.90	£2.00
Cappuccino	£2.10	£2.20
Hot chocolate	£2.00	£2.20
Pot of tea	£1.90	£1.90
Bottled water	£1.50	£1.65

1
1 | What is the new price of 1 cup of hot chocolate?

Correct answer is E: £2.20 – This is an example of a basic interpretation question. You need to look in the 'New price' column to identify the price in the hot chocolate row, which is £2.20.

2 | What is the old price of 1 bottle of water?

Correct answer is A: £1.50 – This is another example of an interpretation question. This time you need to look in the 'Old price' column to identify the price in the bottled water row, which is £1.50.

3 | What is the price of 2 cappuccinos at the old price?

Correct answer is D: £4.20 – This question requires you to look up the correct value for the price of a cappuccino (£2.10) and then multiply it by 2, which gives you the correct answer of £4.20. Beware being caught out by distracters such as 2 cups at the new price (£4.40).

4 | Which drink has the biggest price increase?

Correct answer is C: Hot chocolate – To calculate this answer you need to subtract the old price from the new price for all five items. When you do this you find that the biggest difference in price is of the hot chocolate, which has a difference of 20p between the old price and the new price.

5 | Which drink has the smallest difference between its old and new prices?

Correct answer is D: Pot of tea – This requires the same approach as the previous answer. If you had noted down your calculations for the previous question, when you subtracted the old price from the new price for each item, you would also have the answer for this question already. This is a good example of a time-saving test-taking strategy that may help you to answer more questions correctly within the time limit. There was no difference in the new and old prices of a pot of tea and therefore the answer is D.

How did you do?

Score of 5	Score of 4	Score of 3	Score of 2	Score of 1
Great performance	Good performance	Average performance	Below average	Poor performance

PRODUCTION LINE DATA

A factory has three production lines: Line 1 has 10 workers; Line 2 has 11 workers; and Line 3, which is the Trainee Line, has 10 trainee workers. Each line works 2 sessions per day of 4 hours each. The table below shows the production figures of each line for one particular day.

Line 1 10 workers		Line 2 11 workers		Trainee Line 10 trainee workers	
	Units produced per session		Units produced per session		Units produced per session
Session 1	3,480	Session 1	3,652	Session 1	2,080
Session 2	2,880	Session 2	3,124	Session 2	1,800

How many workers are there in total across the three production lines?

A	B	C	D	E
10	20	31	30	11

2

Calculate the total units produced on this day.

A	B	C	D	E
17,016	13,164	6,360	12,027	15,224

3

What is the difference in the number of units produced by Line 2 in Session 1 and the number of units produced by the same line in Session 2.

A	B	C	D	E
172	3,121	528	631	6,790

4

Calculate the average number of units produced per hour by Line 1 in Session 1.

A	B	C	D	E
2,080	3,480	870	87	875

5

What is the difference between the average number of units produced per hour by Line 2 in Session 1 and Session 2?

A	B	C	D	E
132	917	780	157	848

PRODUCTION LINE DATA

A factory has three production lines: Line 1 has 10 workers; Line 2 has 11 workers; and Line 3, the Trainee Line, has 10 trainee workers. Each line works 2 sessions per day of 4 hours each. The table below shows the production figures of each line for one particular day.

Line 1 10 workers		Line 2 11 workers		Trainee Line 10 trainee workers	
	Units produced per session		Units produced per session		Units produced per session
Session 1	3,480	Session 1	3,652	Session 1	2,080
Session 2	2,880	Session 2	3,124	Session 2	1,800

1   How many workers are there in total across the three production lines?

Correct answer is C: 31 – This question requires you to look up the numbers of workers in each production line in the table and add them together:
Line 1 = **10** workers + Line 2 = **11** workers + Trainee Line = **10** workers = **31** workers

2   Calculate the total units produced on this day.

Correct answer is A: 17,016 – For this question you need to read off the values under the 'Units produced per session' columns for both Sessions 1 and 2 across all three production lines. This produces:
Line 1: 3,480 + 2,880 = 6,360
Line 2: 3,652 + 3,124 = 6,776
Trainee Line: 2,080 + 1,800 = 3,880
Total units produced = 6,360 + 6,776 + 3,880 = **17,016**

This is a more advanced question because it requires two stages of calculation. Note how I have written down my sub-totals in case I need them again later on for another question.

3   What is the difference in the number of units produced by Line 2 in Session 1 and the number of units produced by the same line in Session 2.

Correct answer is C: 528 – We can see from the table that Line 2 produced 3,652 units in Session 1 and 3,124 units in Session 2. To answer the question we therefore need to subtract one value from the other: Difference in units produced: 3,652 – 3,124 = **528**

<table>
<tr><td>**4**</td><td>Calculate Line 1's average number of units produced per hour in Session 1.</td></tr>
</table>

Correct answer is C: 870 – This is a more advanced question again. Firstly you need to review the information presented with the table of data to identify how many hours there are in a session, which is 4 hours. You then need to calculate the average number of units produced by dividing the number of units produced in total during session 1 by 4 (the number of hours in the session):
Units produced in session 1 = 3,480
Divided by the number of hours = 3,480 ÷ 4
Average number of units produced per hour therefore = **870**

<table>
<tr><td>**5**</td><td>What is the difference between Line 2's average units produced per hour in Session 1 and Session 2?</td></tr>
</table>

Correct answer is A: 132 – This question is a progression from the previous one; this time we need to calculate the average number of units produced in both Sessions 1 and 2 for Line 2. We use the same approach as we did in question 4:

Units produced in Session 1 by line 2 = 3,652
Divided by the number of hours = 3,652 ÷ 4
Average number of units produced per hour (Session 1) is therefore = **913**

Units produced in Session 2 by line 2 = 3,124
Divided by the number of hours = 3,124 ÷ 4
Average number of units produced per hour (Session 2) is therefore = **781**

To complete the question we need to subtract the Session 2 figure from the Session 1 figure: Difference between Line 2's average units produced per hour in Session 1 and Session 2: 913 – 781 = **132**

How did you do?

Score of 5	Score of 4	Score of 3	Score of 2	Score of 1
Great performance	Good performance	Average performance	Below average	Poor performance

PRODUCTION LINE DATA

A factory has three production lines: Line 1 has 10 workers; Line 2 has 11 workers; and Line 3, which is the Trainee Line, has 10 trainee workers. Each line works 2 sessions per day of 4 hours each. The table below shows the production figures of each line for one particular day.

Line 1 10 workers		Line 2 11 workers		Trainee Line 10 trainee workers	
	Units produced per session		Units produced per session		Units produced per session
Session 1	3,480	Session 1	3,652	Session 1	2,080
Session 2	2,880	Session 2	3,124	Session 2	1,800

1

If Line 1 had increased its productivity by 10%, how many units would have been produced by Line 1 during the day?

A	B	C	D	E
7,102	7,010	7,156	6,996	6,800

2

What percentage of the day's total production was produced by Line 1?

A	B	C	D	E
2%	37%	26%	56%	30%

3

What is the average number of units produced per worker on the Trainee Line in Session 1?

A	B	C	D	E
3,480	520	180	2,080	208

4

What is the difference in the average number of units produced per day by a trainee worker and a Line 1 worker?

A	B	C	D	E
350	637	386	231	248

5

On the trainee line, what was the average number of units produced per worker per hour during the day?

A	B	C	D	E
52	48.5	970	388	38

PRODUCTION LINE DATA

A factory has three production lines: Line 1 has 10 workers; Line 2 has 11 workers; and Line 3, which is the Trainee Line, has 10 trainee workers. Each line works 2 sessions per day of 4 hours each. The table below shows the production figures of each line for one particular day.

Line 1 10 workers		Line 2 11 workers		Trainee Line 10 trainee workers	
	Units produced per session		Units produced per session		Units produced per session
Session 1	3,480	Session 1	3,652	Session 1	2,080
Session 2	2,880	Session 2	3,124	Session 2	1,800

1   If Line 1 had increased its productivity by 10%, how many units would have been produced by Line 1 during the day?

Correct answer is D: 6,996 – This question requires you to add up the total number of units produced by Line 1 in Sessions 1 and 2: Line 1: 3,480 + 2,880 = 6,360

(You might remember that we arrived at this same subtotal when calculating the answer to Question 1 in Practice test 2 – by recording the subtotal back then we would help speed up our working out in this test.)

We now need to calculate what 10% of the total number of units produced by Line 1 is equal to: 6,360 ÷ 10 = 636

Finally, we need to add this value for 10% to the total number of units produced by Line 1 in both sessions: 6,360 + 636 = **6,996**

2   What percentage of the day's total production was produced by Line 1?

Correct answer is B: 37% – The first step in answering this question is to calculate the total number of units produced by all three lines in both sessions. We calculated this figure when answering Question 2 of Practice test 2:
Line 1: 3,480 + 2,880 = 6,360
Line 2: 3,652 + 3,124 = 6,776
Trainee Line: 2,080 + 1,800 = 3,880

Total units produced = 6,360 + 6,776 + 3,880 = **17,016**

Next we need to know what 1% of this total figure is equal to: 17,016 ÷ 100 = **170.16**

We know from our sub-totals above that Line 1 produced 6,360 units. To work out what percentage this is of the whole day's total we need to divide Line 1's total by the value of 1%:
6,360 ÷ 170.16 = 37.37%

We can see that if we round this figure to the nearest whole number it matches multiple-choice answer **B: 37%**

There is more than one way of calculating this final percentage. Here's an alternative approach:
6,360 × 100 = 636,000
636,000 ÷ 17,016 = 37.37%

3	What is the average number of units produced per worker on the Trainee Line in Session 1?

Correct answer is E: 208 – We can see from the table that the Trainee Line produced 2,080 units in Session 1 and that this line has 10 workers. To answer the question we therefore need to divide the number of units produced by the number of workers:
Average number of units produced per worker on the Trainee Line in Session 1= 2,080 ÷ 10 = **208**

4	What is the difference in the average number of units produced per day by a trainee worker and a Line 1 worker?

Correct answer is E: 248 – This is a progression from the previous question. The first piece of information we need is the average number of units produced per worker for the Trainee Line:

Average number of units produced per worker on the Trainee Line in Sessions 1 and 2: (2,080 + 1,800) ÷ 10 = **388**

We now need to complete the same calculation for Line 1 in Sessions 1 and 2: (3,480 + 2,880) ÷ 10 = **636**

Finally we need to subtract one value from the other: Difference in the average number of units produced per day by a trainee worker and a Line 1 worker: 636 – 388 = **248**

5	On the trainee line, what was the average number of units produced per worker per hour during the day?

Correct answer is B: 48.5 – We already have the first piece of information we need from our calculations in question 4:
Average number of units per worker on the Trainee Line in Sessions 1 and 2 = **388**

The second step is to divide this figure by the number of hours in the day (2 sessions of 4 hours each = 8 hours):
Average number of units produced per worker per hour during the day: 388 ÷ 8 = **48.5**

How did you do?

Score of 5	Score of 4	Score of 3	Score of 2	Score of 1
Great performance	Good performance	Average performance	Below average	Poor performance

CHAPTER 4
NUMERICAL REASONING

Numerical reasoning tests are the most widely used form of numerical ability test because they reflect the level of demands in the majority of jobs. The numerical information in these tests is likely to be more complex than the content of a numeracy test and the answers may require more than one step of calculation.

GETTING THE MOST FROM THESE PRACTICE QUESTIONS

The numerical reasoning practice questions in this chapter are grouped together into 10 short practice tests. Each practice test mirrors the approach of a real test; there is a short extract of numerical information followed by a series of five questions for you to answer about the information you have read.

You're encouraged to complete each practice test one at a time and to spend time at the end of each short test to review your responses against the correct answers. Once you are happy with your work, move on to the next test.

The test instructions for these practice tests follow the same format and approach of instructions in real tests – so to be more prepared for the actual tests, make sure that you familiarise yourself with these.

Completing the practice tests

Ideally you should complete the practice tests in conditions that are as close to the real testing environment and experience as possible.

▸ Find a quiet place that is free from any distractions.
▸ Read the practice test instructions before beginning the first practice test.

▸ Real tests have strict time limits. Simulate this by giving yourself seven minutes per practice test. Start timing from the point you begin reading the numerical information for the test.

▸ Mark your answers for each practice question by circling for the answer you think is correct.

▸ You may use a calculator and rough paper to help with your working (do not use a complex, scientific calculator – only basic pocket calculators are allowed in real tests).

▸ Do not look at the correct answers until you have completed all the practice questions.

Advice for numerical reasoning tests

Numerical reasoning tests are the most commonly used type of numerical test, and you are therefore most likely to encounter this level of test more than any other. When completing a numerical reasoning test, remember these points to perform at your personal best:

▸ **If you have jumped straight to these practice tests** without completing the lower level numeracy tests we encourage you to stop and go back. Completing the numeracy practice tests will greatly benefit your test-taking strategies and level of practice.

▸ **Read the numerical information carefully** before beginning the practice questions. You do not need to memorise it but you will perform better if you are familiar with the information when attempting the questions.

▸ **Never simply write down your answer without checking it first**. Even if you think you have definitely calculated the correct answer, take a moment to check the other choices to make sure you have not been distracted into selecting an incorrect answer.

▶ **Do not simply guess your answers**. You may think that you have a 1 in 5 chance of getting a question correct by guessing, but there is often a scoring mechanism designed to detect or penalise attempts to guess. If you cannot answer a question then review each of the multiple-choice options to identify the ones that are unlikely to be the correct answer. Then you can focus on the shortlist of possible correct answers to identify the one you feel is the right answer.

▶ **Use any time remaining to check your answers**. If you complete all five questions in a practice test within the seven minute time limit, you should re-check your answers as you would in a real test. This is good test-taking strategy and you should aim to make it a habit.

Before you begin the practice tests

Read the test instructions carefully before you begin – these instructions have been adapted from real test instructions. So you will benefit from familiarity with the language and information they include at the time you actually need it.

Reviewing your answers

Once you have completed a practice test you can turn over the page to check your responses against the correct answers. Along with each correct answer we have provided some tips and explanations about the answer.

These practice tests are too short to make an accurate measurement of your numerical reasoning ability, that's what real tests are for, but we can provide some guidance about what your result means:

Score of 5	Score of 4	Score of 3	Score of 2	Score of 1
Great performance	Good performance	Average performance	Below average performance	Poor performance

If you have scored less than what you think your personal best score should be, look at the pattern of your responses – do you tend to answer certain types of question incorrectly? With numerical reasoning tests it is sometimes errors in calculation because of the more difficult nature of some of the questions. Sometimes it is due to particular types of calculation you're less confident about – such as averages or fractions. Or for many people it can be down to misinterpretation of the numerical information when answering the question.

Here's some advice for avoiding these common types of error.

Work the distracters	Remember to examine each of the multiple-choice distracters so you can narrow down the range of answers to just two or three likely candidates
Check the answer against the question	When you have arrived at an answer remember to re-read the question to make sure you have definitely found the result that the question demands
Check your working!	Just because your answer matches one of the multiple-choice answers doesn't guarantee it's correct. Many multiple-choice distracters are based on common calculation errors. Make sure you have avoided any obvious errors and completed all the calculation steps required to answer what the question asks.

Once you have reviewed your answers take a short break before attempting the next test.

NUMERICAL
REASONING
PRACTICE TESTS

PRACTICE TEST INSTRUCTIONS

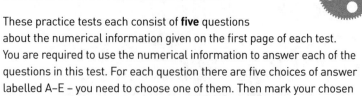

These practice tests each consist of **five** questions
about the numerical information given on the first page of each test.
You are required to use the numerical information to answer each of the
questions in this test. For each question there are five choices of answer
labelled A–E – you need to choose one of them. Then mark your chosen
answer by circling the appropriate option: A, B, C, D or E.

You may use a calculator and rough paper to help with your working-out.

▶ Allow yourself a 7-minute time-limit for each of the practice tests.
▶ You should work quickly and accurately.
▶ If you are not sure of an answer, fill in what you think, but do not
   simply guess your answers.
▶ Check your answers against the correct answers given on the
   pages that follow each practice test.

Remember these important points:

▶ Complete each practice test in a quiet place that is free from any distractions

▶ Start timing the practice test from the point you begin reading the numerical information

▶ Mark your answers for each practice question by circling the answer you think is correct

▶ Remember you can use a pocket calculator and rough paper to help with your answers

▶ Do not turn over to the correct answers until you have completed all the practice questions

Now do the first practice test: start timing yourself and read through the numerical information first. Start answering the questions as soon as you are ready.

CD title	Normal price	Sale price
Pop Factor's Greatest Hits	£13.98	£13.46
Girls Alarmed	£12.50	£10.00
Nose-bleed Techno Mix	£13.49	£12.99
Grime & Punishment	£9.90	£9.38
Old Skool Goth Classics	£18.98	£14.99

1	How much would it cost to buy all 5 CDs at their normal price?			
A £49.87	B £60.82	C £68.85	D £13.77	E £58.85

2	What is the sale discount on *Girls Alarmed* when expressed as a percentage of the normal price?			
A 2.50%	B 2%	C 25%	D 20%	E £2.50

3	A customer buys 4 copies of *Old Skool Goth Classics* in the sale. How much have they saved compared with the normal price?			
A £75.92	B £15.96	C £16.00	D £59.96	E £3.99

4	A customer saves 52p when buying a CD in the sale. Which CD did they buy?			
A *Pop Factor's Greatest Hits*	B *Nose-bleed Techno Mix*	C *Grime & Punishment*	D *Old Skool Goth Classics*	E Cannot say

5	A customer spends £32.44 to buy 1 CD at the normal price and 1 CD at the sale price. Which 2 CDs did they buy?			
A *Old Skool Goth Classics* AND *Pop Factor's Greatest Hits*	B *Girls Alarmed* AND *Grime & Punishment*	C *Pop Factor's Greatest Hits* AND *Nose-bleed Techno Mix*	D *Nose-bleed Techno Mix* AND *Old Skool Goth Classics*	E Cannot say

CD title	Normal price	Sale price
Pop Factor's Greatest Hits	£13.98	£13.46
Girls Alarmed	£12.50	£10.00
Nose-bleed Techno Mix	£13.49	£12.99
Grime & Punishment	£9.90	£9.38
Old Skool Goth Classics	£18.98	£14.99

1   How much would it cost to buy all 5 CDs at their normal price?

Correct answer is C: £68.85 – Some numerical tests, especially power tests, will begin with a gentle question like this to warm you up and help alleviate any nerves rather than hitting you with a difficult one from the off.

This question simply requires you to add the normal prices of the 5 products: £13.98 + £12.50 + £13.49 + £9.90 + £18.98 = **£68.85**

Notice how answer B in the multiple-choice options is £60.82. This is the value you would get if you added the wrong column of prices together (the sale prices). This is a good example of a distracter based on a wrong answer reached by candidates who do not interpret the numerical data correctly.

2   What is the sale discount on *Girls Alarmed* expressed as a percentage of the normal price?

Correct answer is D: 20% – The first step is to calculate the price difference between the normal price and sale price: £12.50 – £10.00 = £2.50

We next need to express this as a percentage of the normal price. One way to do this is to start by working out what 1% is equal to: £12.50÷100 = £0.125

Then we divide the saving in pounds (£2.50) by the value for 1%: £2.50 ÷ 0.125 = **20%**

3   A customer buys 4 copies of *Old Skool Goth Classics* in the sale. How much have they saved compared with the normal price?

Correct answer is B: £15.96 – To answer this question you first need to calculate 2 subtotals; one for the cost of 4 items at the normal price and the second for the cost of 4 items at the sale price:
Normal price: £18.98 × 4 = £75.92 (sub-total 1)
Sale price: £14.99 × 4 = £59.96 (sub-total 2)

Next you need to subtract sub-total 2 from sub-total 1: £75.92 – £59.96 = **£15.96**

Fortunately we are not asked to work out why anybody would want to buy 4 copies of *Old Skool Goth Classics* even if it's in a sale!

4 | A customer saves 52p when buying a CD in the sale. Which CD did they buy?

Correct answer is E: Cannot say – To answer this question you need to subtract the sale price from the normal price of each product and identify the one that has a difference of 52p:
Pop Factor's Greatest Hits: £13.98 – £13.46 = 52p
Girls Alarmed: £12.50 – £10.00 = £2.50
Nose-bleed Techno Mix: £13.49 – £12.99 = 50p
Grime & Punishment: £9.90 – £9.38 = 52p
Old Skool Goth Classics: £18.98 – £14.99 = £3.99

There are a couple of examples of good test-taking strategy in this question. Firstly, you do not need to calculate the difference between normal price and sale price for 2 of the products because they are obviously greater than 52p in difference (Girls Alarmed & Old Skool Goth Classics). This will save you time.

Secondly, if you are checking each of the distracters you'll avoid falling into the trap of selecting the wrong answer of A '*Pop Factor's Greatest Hits*', which has a price differential of 52p. When you check the other answers you see that '*Grime & Punishment*' also has a price differential of 52p and therefore the correct answer is not A (which may have been the first one you found) but E: Cannot say – because you cannot say for certain which product the customer purchased.

5 | A customer spends £32.44 to buy 1 CD at the normal price and 1 CD at the sale price. Which 2 CDs did they buy?

Correct answer is A: *Old Skool Goth Classics* AND *Pop Factor's Greatest Hits* – You could waste a lot of time on this question unless you use good test-taking strategy. If you approach the question by adding together every possible pair of products you could potentially end up performing a few dozen calculations.

A much more effective strategy is to calculate each of the distracters in turn:
Old Skool Goth Classics AND *Pop Factor's Greatest Hits*: £13.98 + £14.99 = 28.97
Old Skool Goth Classics AND *Pop Factor's Greatest Hits*: £13.46 + £18.98 = **32.44**

Rather than answering A straight away it is good strategy to check the other multiple-choice distracters, in case there is a similar trap as in the previous question.
▶ You can see it can't be answer B because neither product is expensive enough to add up to £32.44
▶ It can't be answer C because the price of '*Nose-Bleed Techno Mix*' is different from that of '*Old Skool Goth Classics*'
▶ It can't be answer D because the price of '*Nose-Bleed Techno Mix*' is different from that of '*Pop Factor's Greatest Hits*'

There are other ways to arrive at this answer – but some are going to be more time efficient than others. It's helpful to look for shortcuts that you can safely use when completing numerical tests – the time you save may enable you to attempt more questions within the time limit. This is partly about good test-taking strategy but it also relies on your true numerical ability – the stronger your ability the easier you will find it to quickly discount some multiple-choice distracters using rough mental arithmetic estimation rather than more long-winded calculations.

How did you do?

Score of 5	Score of 4	Score of 3	Score of 2	Score of 1
Great performance	Good performance	Average performance	Below average	Poor performance

Products	Selling Price	Profit
Hand towels	£4.99	£1.00
Bath towels	£6.64	£1.98
Face flannels	£1.49	66p
Bath sheets	£8.99	£1.35
Bath mats	£3.49	28p

1	If profit margin is calculated as the profit divided by the selling price expressed as a percentage, which item gives the greatest profit margin?

A	B	C	D	E
Hand towels	Bath towels	Face flannels	Bath sheets	Bath mats

2	The store sells 35 of the same product and makes £47.25 profit. Which item did it sell?

A	B	C	D	E
Hand towels	Bath towels	Face flannels	Bath sheets	Bath mats

3	What is the average profit across the 5 products (rounded to the nearest penny)?

A	B	C	D	E
£19.66	£5.27	£1.50	£1.05	£5.12

4	If the store sells 18 Hand towels and 103 Bath mats in a week how much profit will it make from these sales?

A	B	C	D	E
£28.84	£46.84	£449.29	£64.48	Cannot say

5	How many Face flannels must the store sell in order to make the same profit as made when selling 50 Bath towels?

A	B	C	D	E
1500	102	95	17	150

Products	Selling Price	Profit
Hand towels	£4.99	£1.00
Bath towels	£6.64	£1.98
Face flannels	£1.49	66p
Bath sheets	£8.99	£1.35
Bath mats	£3.49	28p

1 If profit margin is calculated as the profit divided by the selling price expressed as a percentage, which item gives the greatest profit margin?

Correct answer is C: Face flannels – This is an example of a question that explains a term (in this case profit) for the benefit of candidates who may be less familiar with the term than others. In this way the test levels the playing field and ensures no one is disadvantaged for social, cultural or educational reasons.

One way of completing this question is to simply divide the profit by the selling price and then multiply it by 100 to express it as a percentage, as the question suggests:
Hand towels: £1.00 ÷ £4.99 × 100 = 20%
Bath towels: £1.98 ÷ £6.64 × 100 = 29.81%
Face flannels: £0.66 ÷ £1.49 × 100 = **44.2%**
Bath sheets: £1.35 ÷ £8.99 × 100 = 15%
Bath mats: £0.28 ÷ £3.49 × 100 = 8%

The correct answer is therefore **Face flannels**.

2 The store sells 35 of the same product and makes £47.25 profit. Which item did it sell?

Correct answer is D: Bath sheets – You need to divide the profit by the number of items sold: £47.25 ÷ 35 = **£1.35**

You can see from the table that the only product with a **£1.35** profit value is **Bath sheets**, which is therefore the correct answer.

3 What is the average profit across the 5 products (rounded to the nearest penny)?

Correct answer is D: £1.05 – To calculate the average profit you need to add together the profit figures for each product and then divide that sum by the number of products (5):
£1.00 + £1.98 + 66p + £1.35 + 28p = £5.27
£5.27 ÷ 5 = £1.054

Therefore, the average profit across the five products, rounded to the nearest penny, is **£1.05**

4

If the store sells 18 Hand towels and 103 Bath mats in a week how much profit will it make from these sales?

Correct answer is B: £46.84 – To answer this question you need to multiply the profit figures for each of the 2 products by the quantity sold and then add these 2 sub-totals together:
Hand towels: £1.00 × 18 = £18.00
Bath mats: 28p × 103 = £28.84

Total profit = £18.00 + £28.84 = **£46.84**

5

How many Face flannels must the store sell in order to make the same profit as made when selling 50 Bath towels?

Correct answer is E: 150 – If you divide the profit figure for Bath towels by the profit figure of the Face flannel you identify how many times more profitable the Bath towels are per unit:
£1.98 ÷ £0.66 = £3

So, for every Bath towel sold you need to sell 3 Face flannels to make the same profit. If 50 Bath towels have been sold you would need to sell 3 times this number of Face flannels to make the equivalent amount of profit: 3 × 50 = 150

The answer is therefore **150** Face flannels.

There's an alternative method for answering this question where you calculate how much profit 50 Bath towels generates and then divide that result by the profit figure for Face flannels. Go ahead and try it.

How did you do?

Score of 5	Score of 4	Score of 3	Score of 2	Score of 1
Great performance	Good performance	Average performance	Below average	Poor performance

During an average day, 10,000 different people visit *FairShare* Website.
Details were taken of the first 2,500 visitors for use in this survey.

	Up to 12 Years	3%
	13 - 18 Years	27%
	19 - 24 Years	29%
	25 - 30 Years	22%
	31 - 36 Years	12%
	37 - 42 Years	5%
	43 + Years	2%

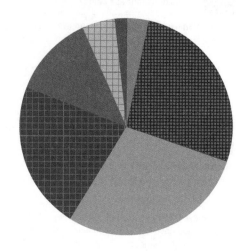

1	How many of the 2,500 surveyed visitors were 24 years old and under?				
A	B	C	D	E
59	725	1,475	2,500	5,900

2	If the volume of visitors at weekends increases by 15%, how many visitors in the 37–42 years age range would you expect on a Saturday?				
A	B	C	D	E
5	20	125	144	575

3	Assuming the visits are evenly distributed throughout a 12-hour day, how many would you expect to have been received between 6 p.m. and 8 p.m. on a typical day?				
A	B	C	D	E
208	417	833	1,667	2,500

4	On a day when there were 12,600 visits, which age group would you expect to account for 3,654 visits?				
A	B	C	D	E
Up to 12 years	13–18 years	19–24 years	25–30 years	31–36 years

5	On 1 particular day, there were 3 times the usual number of visitors to the website. Estimate how many of these would have been aged between 19 and 36 years.				
A	B	C	D	E
63	3,600	18,900	30,000	1,890,000

During an average day, 10,000 different people visit *FairShare* Website.
Details were taken of the first 2,500 visitors for use in this survey.

Up to 12 Years	3%
13 - 18 Years	27%
19 - 24 Years	29%
25 - 30 Years	22%
31 - 36 Years	12%
37 - 42 Years	5%
43 + Years	2%

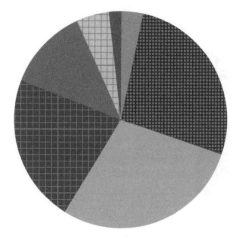

---

1   How many of the 2,500 surveyed visitors were 24 years old and under?

Correct answer is C: 1,475 – You can see from the pie chart that the visitors under the age of 24 lie in the first 3 categories. You therefore need to add the percentages of these three groups: 3 + 27 + 29 = 59%

This is not the final answer (though it is the first multiple-choice distracter); you are asked for the number of visitors who are 24 and under, not the percentage. You therefore now need to calculate how many visitors 59% of 2,500 people is equal to: 2,500 × 59 ÷ 100 = **1,475** visitors

---

2   If the volume of visitors at weekends increases by 15%, how many visitors in the 37–42 years age range would you expect on a Saturday?

Correct answer is E: 575 – To start your answer you need to calculate what an increase of 15% means for the total number of visitors, which is usually 10,000: 10,000 × 15 ÷ 100 = 1,500 extra visitors

The total number of visitors you'd expect per day at the weekend is therefore 10,000 + 1,500 = 11,500

We can see from the pie chart that the 37–42 years age range accounts for 5% of visitors. We therefore need to calculate what 5% of 11,500 people is equal to: 11,500 × 5 ÷ 100 = 575

A common mistake with this question is to base your answer on the 2,500 figure – which is the survey sample size, not the number of daily visitors. Multiple-choice distracter D is the wrong answer at which you'd arrive.

3	Assuming the visits are evenly distributed throughout a 12-hour day, how many would you expect to have been received between 6 p.m. and 8 p.m. on a typical day?

Correct answer is D: 1,667 – If the usual number of visits in a 12-hour day is 10,000, then the number of visits per hour must equal this number divided by 12: $10,000 \div 12 = 833.33$

This is not yet the right answer since you are being asked for visitors received in a period of 2 hours (between 6p.m. and 8 p.m.). The correct answer is therefore: $833.33 \times 2 = 1,666.67$

Which rounded to a whole number gives **1,667**

4	On a day when there were 12,600 visits, which age group would you expect to account for 3,654 visits?

Correct answer is C: 19 – 24 years – The quickest way to do this is to calculate what percentage of 12,600 is 3,654 equal to: $3,654 \div 12,600 \times 100 = 29\%$

The only age group that accounts for 29% of visits is the **19–24 years** age group; this is therefore the correct answer.

5	On 1 particular day, there were 3 times the usual number of visitors to the website. Estimate how many of these would have been aged between 19 and 36 years.

Correct answer is C: 18,900 – 3 times the usual number of visitors would be 30,000 (10,000 $\times$ 3). The percentage of visitors in the 19–36 age range is 63% (you need to add together the middle 3 age ranges).

You then need to calculate what 63% of 30,000 is equal to: $30,000 \times 63 \div 100 = \mathbf{18,900}$

How did you do?

Score of 5	Score of 4	Score of 3	Score of 2	Score of 1
Great performance	Good performance	Average performance	Below average	Poor performance

NUMBER OF NEW FILES UPLOADED EACH QUARTER TO THE FAIRSHARE WEBSITE

Quarter	Audio files added	Video files added	Image files added	Text files added	Total files added
1st quarter (Jan, Feb, Mar)	2,860	1,104	1,680	102	5,746
2nd quarter (Apr, May, Jun)	12,980	3,690	8,650	380	25,700
3rd quarter (Jul, Aug, Sept)	107,570	16,570	32,060	2,200	158,400
4th quarter (Oct, Nov, Dec)	401,128	32,046	86,540	7,980	527,694

1 What percentage of the total number of files added in the 4th quarter were audio files?

A	B	C	D	E
1.32%	68%	76%	132%	40,1128

2 How many image and text files were added in the second half of the year?

A	B	C	D	E
10,180	32,280	94,520	118,600	128,780

3 What is the difference between the average number of audio files added in the first half of the year and those added in the second half of the year?

A	B	C	D	E
7,920	15,840	246,429	254,349	262,269

4 Express as a percentage how many more video files than text files were added in the 1st quarter.

A	B	C	D	E
83%	102%	1002%	982.35%	1,182%

5 Express the number of audio files added in the 2nd quarter as a proportion of the total number of files added during the same quarter.

A	B	C	D	E
1/4	1/3	1/2	2/3	3/4

NUMBER OF NEW FILES UPLOADED EACH QUARTER TO THE FAIRSHARE WEBSITE

Quarter	Audio files added	Video files added	Image files added	Text files added	Total files added
1st quarter (Jan, Feb, Mar)	2,860	1,104	1,680	102	5,746
2nd quarter (Apr, May, Jun)	12,980	3,690	8,650	380	25,700
3rd quarter (Jul, Aug, Sept)	107,570	16,570	32,060	2,200	158,400
4th quarter (Oct, Nov, Dec)	401,128	32,046	86,540	7,980	527,694

1   What percentage of the total number of files added in the 4th quarter were audio files?

Correct answer is C: 76% – For this question you need to identify the correct values from the table; the number of audio files added in the 4th quarter and the total files added in the same period. You should then use these figures to calculate the percentage: 401,128 ÷ 527,694 × 100 = 76.01

Therefore: that **76%** of the total files added in the 4th quarter were audio files.

2   How many image and text files were added in the second half of the year?

Correct answer is E: 128,780 – The second half of the year is the period July to December. You therefore need to add the 4 values for image and text files from this period together to arrive at the correct answer: 32,060 + 2,200 + 86,540 + 7,980 = 128,780

Therefore **128,780** image and text files were added in the second half of the year.

<table>
<tr><td>**3**</td><td>What is the difference between the average number of audio files added in the first half of the year and those added in the second half of the year?</td></tr>
</table>

Correct answer is C: 246,429 – You need to begin by calculating the 2 averages: 1 for the average number of audio files added in the first half of the year and the second for those added in the second half of the year:
Average number of audio files for January to June: 2,860 + 12,980 ÷ 2 = 7,920
Average number of audio files for July to December: 107,570 + 401,128 ÷ by 2 = 254,349

You then need to subtract one average from the other to find the difference: 254,349 – 7,920 = 246,429

The difference between the average number of audio files added in the first half of the year and those added in the second half of the year is therefore = **246,429**

<table>
<tr><td>**4**</td><td>Express as a percentage how many more video files than text files were added in the first quarter.</td></tr>
</table>

Correct answer is D: 982.35% – This question turns up the heat a little. The first step is to calculate the difference in numerical (rather than percentage) terms: Video files added in 1st quarter = 1,104 – text files added in 1st quarter = 102

The difference is therefore 1,002

You then need to convert this into the percentage: 1,002 ÷ 102 × 100 = 982.35%

Therefore the proportion of video files added in the 1st quarter was **982.35%** greater than the number of text files added.

<table>
<tr><td>**5**</td><td>Express the number of audio files added in the 2nd quarter as a proportion of the total number of files added during the same quarter.</td></tr>
</table>

Correct answer is C: $\frac{1}{2}$ – You need to identify 2 sub-totals to begin with: the number of audio files and the total number of files added in the 2nd quarter:
Audio files added in 2nd quarter: 12,980
Total number of files added in 2nd quarter: 25,700

There are two ways of identifying the proportion: One is to divide the total number of files added by the number of audio files: 25,700 ÷ 12,980 = 1.97 – which indicates that the number of audio files added was about half the number of all files. Another way is to calculate the proportion as a percentage: 12,980 ÷ 25,700 × 100 = 50.50%, which again is around one half.

Both answers match the answer of C: $\frac{1}{2}$

How did you do?

Score of 5	Score of 4	Score of 3	Score of 2	Score of 1
Great performance	Good performance	Average performance	Below average	Poor performance

FAIRSHARE MERCHANDISE

	T-shirt	Baseball cap	Mouse mat	Hoodie	Mug
Item **cost** to FairShare	£8.45	£8.95	£5.15	£14.95	£2.90
Item **selling price** to customer	£11.99	£12.95	£7.99	£18.45	£5.22
Number sold last month	69	18	88	22	41

1	What is the cost of 20 Baseball caps?				
	A £8.95	B £12.95	C £80	D £179	E £259

2	If profit is defined as the selling price minus the cost, how much profit was generated by the Hoodies in the last month?				
	A £3.50	B £22	C £77	D £328.90	E £405.90

3	Which of the 5 items of merchandise has the smallest percentage mark-up on selling price?				
	A Baseball cap	B Mouse mat	C Hoodie	D Mug	E Cannot say

4	What percentage of the selling price to the customer is the cost to FairShare for the Mouse mat?				
	A 0.41%	B 35.55%	C 41.14%	D 64.45%	E 155%

5	A product increased its sales by approximately 11% in the next month and sold an extra 2 items. Which product was it?				
	A T-Shirt	B Baseball cap	C Hoodie	D Mug	E Cannot say

FAIRSHARE MERCHANDISE

	T-shirt	Baseball cap	Mouse mat	Hoodie	Mug
Item **cost** to FairShare	£8.45	£8.95	£5.15	£14.95	£2.90
Item **selling price** to customer	£11.99	£12.95	£7.99	£18.45	£5.22
Number sold last month	69	18	88	22	41

1 What is the cost of 20 Baseball caps?

Correct answer is D: £179 – This is a relatively straightforward warm-up question. You need to multiply the cost price of the baseball cap by 20: £8.95 × 20 = 179

Therefore the cost of 20 Baseball caps is **£179**

2 If profit is defined as the selling price minus the cost, how much profit was generated by the Hoodie in the last month?

Correct answer is C: £77 – The definition of profit is provided to ensure no candidates are disadvantaged through being unfamiliar with this term. To answer the question you need to follow this definition:
Hoodie selling price minus hoodie cost price: £18.45 – £14.95 = £3.50

Don't stop there – you haven't yet answered the question (despite this value matching the first multiple-choice distracter). You need to calculate what this figure means in terms of the profit for the month. This means multiplying it by the number of hoodies sold: 3.5 × 22 = 77

Therefore hoodies generated £77 of profit in the last month.

3	Which of the five items of merchandise has the smallest percentage mark-up on selling price?

Correct answer is C: Hoodie – The first step is to work out the profit margins/mark-ups for the 4 products (remember that the T-shirt was not an option in the answers to this question):
Baseball Cap: £12.95 – £8.95 = £4.00
Mouse mat: £7.99 – £5.15 = £2.84
Hoodie: £18.45 – £14.95 = £3.50
Mug: £5.22 – £2.90 = £2.32

You then need to convert these into percentages by dividing the mark-up figure by the selling cost figure and multiplying by 100:
Baseball Cap: $4.00 \div 12.95 \times 100 = 30.88\%$
Mouse mat: $2.84 \div 7.99 \times 100 = 35.54\%$
Hoodie: $3.50 \div 18.45 \times 100 = 18.97\%$
Mug: $2.32 \div 5.22 \times 100 = 44.44\%$

The correct answer is therefore the **Hoodie** since its profit margin is the lowest in percentage terms at 18.97%

4	What percentage of the selling price to the customer is the cost to FairShare for the Mouse mat?

Correct answer is D: 64.45% – To work this out you must simply calculate the percentage of the selling price that the cost is equal to: $5.15 \div 7.99 \times 100 = 64.45$

Therefore the percentage of the selling price to the customer of the cost to FairShare for the Mouse mat is **64.45%**

5	A product increased its sales by approximately 11% in the next month and sold an extra 2 items. Which product was it?

Correct answer is E: Cannot say – You might be able to use your mental arithmetic skills to exclude any products that are obviously wrong answers by quickly calculating what 10% of their number sold last month is equal to, and seeing if that equates to an extra 2 items. Alternatively you can calculate what 11% of each product's number sold last month is equal to (note that the Mouse mat was not an option in the answers to this question):

T-Shirt: $69 \div 100 \times 11 = 7.59$ – approximately 8 items
Baseball Cap: $18 \div 100 \times 11 = 1.98$ – **approximately 2 items**
Hoodie: $22 \div 100 \times 11 = 2.42$ – **approximately 2 items**
Mug: $41 \div 100 \times 11 = 4.51$ – approximately 5 items

The answer is therefore cannot say because if both the Baseball cap and Hoodie sold 2 more items, the percentage increase for both approximates to 11%

How did you do?

Score of 5	Score of 4	Score of 3	Score of 2	Score of 1
Great performance	Good performance	Average performance	Below average	Poor performance

Customer telephone bill summaries for September		
Customer name	**Total telephone calls (minutes)**	**Call plan**
Jamie McCrimmon	392	Easycall
Jo Grant	411	Easycall
Harry Sullivan	297	Standard
Tegan Jovanka	105	Low-use
Ben Jackson	167	Premium
Susan Foreman	532	Premium

Call plan costs		
Name of call plan	**Cost of calls (per minute)**	**Monthly line charge**
Standard	£0.05	£11.99
Premium	£0.04	£12.50
Easycall	£0.03	£15.00
Low-use	£0.10	£7.99
Anytime	No charge	£25.00

1	Calculate Harry Sullivan's September telephone bill total.				
	A £11.99	B £23.91	C £26.84	D £297	E £3,561.03

2	Estimate Jo Grant's annual telephone bill based on her September bill total.				
	A £12.33	B £27.33	C £147.96	D £180	E £327.96

3	Susan Foreman receives a 7.5% discount on her telephone bill because she pays by direct debit. What is Susan's September bill after this discount?				
	A £3.19	B £28.90	C £31.25	D £39.37	E £42.56

4	If Harry Sullivan changed his telephone call plan to include an extra line and pays an additional £7.99 monthly charge, what would be his September bill when including a 7.5% direct debit discount?				
	A £32.22	B £32.82	C £34.83	D £37.44	E £39.44

5	Which customer would have saved the most money in September if they were on the Anytime call plan?				
	A Jamie McCrimmon	B Jo Grant	C Tegan Jovanka	D Susan Foreman	E Cannot say

Customer telephone bill summaries for September		
Customer name	**Total telephone calls (minutes)**	**Call plan**
Jamie McCrimmon	392	Easycall
Jo Grant	411	Easycall
Harry Sullivan	297	Standard
Tegan Jovanka	105	Low-use
Ben Jackson	167	Premium
Susan Foreman	532	Premium

Call plan costs		
Name of call plan	**Cost of calls (per minute)**	**Monthly line charge**
Standard	£0.05	£11.99
Premium	£0.04	£12.50
Easycall	£0.03	£15.00
Low-use	£0.10	£7.99
Anytime	No charge	£25.00

1   Calculate Harry Sullivan's September telephone bill total.

Correct answer is C: £26.84 – You can see from the first table that Harry is on the Standard call plan and has used 297 minutes in September. To calculate his bill for September you need to identify what the cost per minute is for telephone calls on his call plan and multiply this by the minutes Harry has used: Standard call plan: £0.05 × 297 minutes = £14.85

Don't stop there! You need to also include the monthly line charge for Harry's call plan: Standard call plan monthly charge: £11.99 + £14.85 = £26.84

Therefore Harry's September bill total is **£26.84**

2   Estimate Jo Grant's annual telephone bill based on her September bill total.

Correct answer is E: £327.96 – The first step is to calculate Jo's September bill total:
Easycall call plan: £0.03 × 411 minutes = £12.33
Easycall call plan monthly charge: £15.00 + £12.33 = £27.33

You can use this September total as a basis for estimating Jo's annual bill by multiplying it by 12: £27.33 × 12 = £327.96

Therefore, Jo Grant's total annual telephone bill estimated from her September bill total is **£327.96**

3

Susan Foreman receives a 7.5% discount on her telephone bill because she pays by direct debit. What is Susan's September bill after this discount?

Correct answer is C: £31.25 – The first step is to calculate Susan's September bill total:
Premium call plan: £0.04 × 532 minutes = £21.28
Premium call plan monthly charge: £12.50 + £21.28 = £33.78

Next you need to calculate what 7.5% of the total bill is equal to: £33.78 ÷ 100 × 7.5 = £2.53

Finally, you need to subtract this discount from the total bill: £33.78 – £2.53 = £31.25

Therefore, Susan's recalculated September bill total, to include the direct debit discount, is **£31.25**

4

If Harry Sullivan changed his telephone call plan to include an extra line and pays an additional £7.99 monthly charge, what would be his September bill when including a 7.5% direct debit discount?

Correct answer is A: £32.22 – You already have the first piece of information you need from your answer to question 1; Harry's September bill total of £26.84. To answer this question you need to add the additional cost of the extra line and apply the direct debit discount: Extra line: £26.84 + £7.99 = £34.83. Discount for direct debit = £34.83 ÷ 100 × 7.5 = £2.61

Finally you need to subtract this discount from the bill total: £34.83 – £2.61 = £32.22

Therefore, Harry's recalculated September bill total, to include the extra line charge and direct debit discount, is **£32.22**

5

Which customer would have saved the most money in September if they were on the Anytime call plan?

Correct answer is D: Susan Foreman – First, you need to calculate each customer's September bill:
Jamie McCrimmon: 392 × £0.03 + £15.00 = £26.76
Jo Grant: 411 × £0.03 + £15.00 = £27.33
Tegan Jovanka: 105 × £0.10 + £7.99 = £18.49
Susan Foreman: 532 × £0.04 + £12.50 = £33.78

The cost of the Anytime call plan is £25 per month with no charge for calls, so the customer who would have saved the most money in September if they'd been on this call plan is Susan Foreman.

How did you do?

Score of 5	Score of 4	Score of 3	Score of 2	Score of 1
Great performance	Good performance	Average performance	Below average	Poor performance

Energy saving methods	Typical cost	Typical annual saving
Low-energy light bulbs	£45.00	£9.00
Cavity wall insulation	£550	£115
Draught proofing	£25	£9.95
Roof insulation	£200	£7.50
Double glazing	£2,000	£25.00

1	What is the average cost of the five energy saving methods?			
A £33.29	B £564	C £707.25	D £2,820	E £4,100

2	How many years would it take to break even on the typical cost of cavity wall insulation?			
A 5 years	B 10 years	C 15 years	D 20 years	E 25 years

3	If a household invested in low-energy light bulbs and draught proofing how much would it save over 15 years?			
A £18.95	B £70	C £214.25	D £284.25	E £765.75

4	Which energy saving method saves the most money over 10 years?			
A Low-energy light bulbs	B Cavity wall insulation	C Draught proofing	D Roof insulation	E Double glazing

5	If a typical household installed roof insulation, what percentage of the cost would it have saved in energy costs over 5 years?			
A 5.33%	B 18.75%	C 37.5%	D 50%	E 75%

PRACTICE TEST 7: REVIEW YOUR ANSWERS

Energy saving methods	Typical cost	Typical annual saving
Low-energy light bulbs	£45.00	£9.00
Cavity wall insulation	£550	£115
Draught proofing	£25	£9.95
Roof insulation	£200	£7.50
Double glazing	£2,000	£25.00

1 What is the average cost of the five energy saving methods?

Correct answer is B: £564 – You need to add the typical costs of the 5 energy saving methods and divide the sum by their number (5): £45.00 + £550 + £25 + £200 + £2,000 = £2,820. £2,820 ÷ 5 = £564

Therefore, the average cost of the four energy saving methods is **£564**

2 How many years would it take to break even on the typical cost of cavity wall insulation?

Correct answer is A: 5 years – You need to divide the typical cost by the annual saving: £550 ÷ £115 = 4.78

Therefore, it would take 5 years to break even on the typical cost of cavity wall insulation.

3 If a household invested in low-energy light bulbs and draught proofing how much would it save over 15 years?

Correct answer is C: £214.25 – The first step is to add together the annual saving of each of the 2 methods and multiply this by 15 years: £9.00 + £9.95 = £18.95. £18.95 × 15 = £284.25

Don't forget to subtract the original cost of installing the 2 methods: £45 + £25 = £70 £284.25 – £70 = £214.25

Therefore, a household that invested in low-energy light bulbs and draught proofing would save **£214.25** over 15 years.

4   **Which energy saving method saves the most money over 10 years?**

Correct answer is B: Cavity wall insulation – You might be able to rule out some of the energy saving methods using mental arithmetic alone. Alternatively, you need to calculate the saving for each method, remembering to subtract the installation cost of each at the same time:

Low-energy light bulbs: £9 × 10 – £45 = £45
Cavity wall insulation: £115 × 10 – £550 = £600
Draught proofing: £9.95 × 10 – £25 = £74.50
Roof insulation: £7.50 × 10 – £200 = –£125
Double glazing: £25 × 10 – £2,000 = –£1,750

The energy saving method that saves the most money over 10 years is therefore **Cavity wall insulation**.

5   **If a typical household installed roof insulation, what percentage of the cost would they have saved in energy costs over 5 years?**

Correct answer is B: 18.75% – The first step is to calculate the saving after 5 years: £7.50 × 5 = £37.50

You now need to express this as a percentage: 37.50 ÷ 200 × 100 = 18.75

Therefore the household would recover **18.75%** of the cost of installing roof insulation over 5 years.

How did you do?

Score of 5	Score of 4	Score of 3	Score of 2	Score of 1
Great performance	Good performance	Average performance	Below average	Poor performance

CUSTOMER SATISFACTION SURVEY RESULTS

Number of customers surveyed = **200**

Rating area	Approval rating (% good– excellent); target 80%	Improvement on last survey (%)
Tidiness of store environment	82%	+1%
Availability of products	79%	–2%
Quality of deli staff service	84%	+5%
Service in customer restaurant	82%	–1%
Helpfulness of store staff	80%	–1%
Knowledge of store staff	78%	–3%
Performance of staff at check-out	78%	+1%
Would shop at the store again	81%	0%
Overall customer satisfaction	80%	+2%

1	Which rating area had the biggest percentage improvement compared with the last survey?				
A Tidiness of store environment	B Availability of products	C Quality of deli staff service	D Service in customer restaurant	E Cannot say

2	In which rating area did the store perform most badly against its target?				
A Service in customer restaurant	B Helpfulness of store staff	C Knowledge of store staff	D Performance of staff at check-out	E Cannot say

3	How many more customers needed to rate 'Availability of products' as good-excellent for the store to beat its target by 5%?				
A 2	B 4	C 6	D 10	E 12

4	If 18 more customers respond to the next survey, what percentage increase will this represent on the number of customers surveyed this time?				
A 4.5%	B 9%	C 18%	D 36%	E Cannot say

5	What is the average percentage difference over the 9 rating areas since the last survey?				
A +0.22%	B +0.25%	C +0.44%	D 4%	E 80.44%

CUSTOMER SATISFACTION SURVEY RESULTS

Number of customers surveyed = **200**

Rating area	Approval rating (% good-excellent); target 80%	Improvement on last survey (%)
Tidiness of store environment	82%	+1%
Availability of products	79%	−2%
Quality of deli staff service	84%	+5%
Service in customer restaurant	82%	−1%
Helpfulness of store staff	80%	−1%
Knowledge of store staff	78%	−3%
Performance of staff at check-out	78%	+1%
Would shop at the store again	81%	0%
Overall customer satisfaction	80%	+2%

1 Which rating area had the biggest percentage improvement compared with the last survey?

Correct answer is C: Quality of deli staff service – This is a simple interpretation question – do not be fooled into embarking on performing complex calculations; you simply need to look up the correct answer in the third column. The rating area with the biggest percentage improvement is the 'Quality of deli staff service', with a 5% increase. Perhaps the deli counter changed the colour of the queue number tickets to a more pleasing hue.

This is an example of a question you sometimes get in the early stages of a numerical reasoning testing to check you can interpret the provided numerical information correctly and to demonstrate how a table or graph works.

2 In which rating area did the store perform most badly against its target?

Correct answer is E: Cannot say – At first glance this may seem like another straightforward interpretation question, if you quickly scan the table you can see that the 'Knowledge of store staff' rating area scored 78%, which looks like the lowest. But when you check more carefully you can see that another rating area, 'Performance of staff at check-out' also scored 78%. We therefore cannot say for certain which single rating area performed most badly since both had the same target (80%). The correct answer is therefore **Cannot say**.

| **3** | **How many more customers needed to rate 'Availability of products' as good-excellent for the store to beat its target by 5%?** |

Correct answer is E: 12 – To beat the 80% target by 5% in this rating area the store needs to score another 6% (it is currently at 79% and would need to move to 85%). You need to identify what this 6% figure equates to in terms of actual number of customers. Since 200 customers were surveyed, this is the figure to base your answer on: 6 × 200 ÷ 100 = 12

Therefore, **12 more people** needed to rate 'Availability of products' as good-excellent for the store to achieve 85% and thus beat its target by 5%

| **4** | **If 18 more customers respond to the next survey, what percentage increase will this represent on the number of customers surveyed this time?** |

Correct answer is B: 9% – 200 people completed the survey; we therefore need to calculate what 18 people equates to in percentage terms based on this total: 18 ÷ 200 × 100 = 9%

Therefore an increase in customers surveyed from 200 to 218 represents a percentage increase of **9%**

| **5** | **What is the average percentage difference over the 9 rating areas since the last survey?** |

Correct answer is A: +0.22% – You need to add together the 9 values in the column on the right and divide by 9 to get the average (watch out for the fiddly minus signs):
1 + (–2) + 5 + (–1) + (–1) + (–3) + 1 + 0 + 2 = 2
2 ÷ 9 = 0.22

Therefore, the average percentage difference over the 9 rating areas since the last survey is **+0.22%**

How did you do?

Score of 5	Score of 4	Score of 3	Score of 2	Score of 1
Great performance	Good performance	Average performance	Below average	Poor performance

FAIRSHARE WEBSITE ADVERTISING PRICES

Type of advert	Cost of 1 impression (advert is displayed to 1,000 users)	Cost per click (each time advert is clicked on by user)
Large banner	£12	£0.20
Small banner	£10	£0.16
Long video	£15	£0.35
Short video	£12	£0.31
Pop-up advert	£7.50	£0.20

1 How much would it cost for a Large banner advert to be shown 2,000 times and clicked on 5 times?

A	B	C	D	E
£13	£25	£28	£241	£460

2 What is the average cost per click price for all types of advert?

A	B	C	D	E
16p	20p	24p	31p	£1.22

3 An advertiser buys a Small banner advert. Of the 1,000 people who see it, only 3% click on it and buy the product advertised. How much does each sale cost the advertiser?

A	B	C	D	E
16p	49p	£3.40	£4.80	£14.80

4 What percentage of the cost of 1 impression of the Large banner advert is the cost per click?

A	B	C	D	E
1.67%	6%	60%	167%	240%

5 If a client paid £14.17 for 1 impression and 7 clicks, which type of advert did they choose?

A	B	C	D	E
Large banner	Small banner	Long video	Short video	Pop-up advert

FAIRSHARE WEBSITE ADVERTISING PRICES

Type of advert	Cost of 1 impression (advert is displayed to 1,000 users)	Cost per click (Each time advert is clicked on by user)
Large banner	£12	£0.20
Small banner	£10	£0.16
Long video	£15	£0.35
Short video	£12	£0.31
Pop-up advert	£7.50	£0.20

1 How much would it cost for a Large banner advert to be shown 2,000 times and clicked on 5 times?

Correct answer is B: £25 – This question requires 2 calculations: the cost of 2,000 impressions and the cost of 5 clicks: Cost of 2,000 impressions: £12 × 2 = £24. Cost of 5 clicks: £0.20 × 5 = £1.00

To complete your answer you need to add these sub-totals together: £24 + £1 = £25

Therefore the cost for a Large banner advert to be shown 2,000 times and clicked on 5 times is £25

2 What is the average cost per click price for all types of advert?

Correct answer is C: 24p – This simply requires you to calculate the average cost by adding the 5 costs together and dividing by their number (5): £0.20 + £0.16 + £0.35 + £0.31 + £0.20 = £1.22. £1.22 ÷ 5 = £0.244

Therefore the average cost per click price for all advertisement types is 24p

3 An advertiser buys a Small banner advert. Of the 1,000 people who see it, only 3% click on it and buy the product advertised. How much does each sale cost the advertiser?

Correct answer is B: 49p – The advertiser has paid for 1,000 impressions, which, as you can see from the table, costs £10. You can also see from the table that a click on the small banner ad costs 16p

You first need to calculate how many people clicked on the advert. This means working out what 3% of 1,000 people is equal to: $3 \times 1,000 \div 100 = 30$ people

All 30 people clicked on the advert before buying the product. We know each click costs 16p, therefore the total cost of the click-throughs = $30 \times £0.16 = £4.80$

Therefore the total cost to the advertiser is the cost of the advert plus the cost of these clicks-throughs: £10 + £4.80 = £14.80

That is not the end of the question (don't be tempted by the incorrect distracter). The question asks you how much **each** sale costs the advertiser – we therefore need to divide this total cost by the number of sales (30 people): $£14.80 \div 30 = £0.49$

Therefore each sale costs the advertiser **49p**

4 What percentage of the cost of 1 impression of the Large banner advert is the cost per click?

Correct answer is A: 1.67% – The cost of 1 impression of the Large banner advert is £12. The cost of a click on a Large banner advert is 20p. You therefore need to express this as a percentage: $0.20 \div 12 \times 100 = 1.67\%$

Therefore, the percentage of the cost per click of one impression of the Large banner advert is **1.67%**

5 If a client paid £14.17 for 1 impression and 7 clicks, which type of advert did they choose?

Correct answer is D: Short video – The first step is to exclude any obviously wrong distracters; the correct answer cannot be C: Long video because the cost of 1 impression is £15, which is greater than £14.17. You can then work through the remaining options by calculating the cost of 7 clicks for each and adding this to the cost of 1 impression:
Large banner: $7 \times £0.20 = £1.40 + £12 = £13.40$
Small banner: $7 \times £0.16 = £1.12 + £10 = £11.20$
Short video : $7 \times £0.31 = £2.17 + £12 = £14.17$
Pop-up advert: $7 \times £0.20 = £1.40 + £7.50 = £8.90$

The correct answer is therefore **Short video**.

How did you do?

Score of 5	Score of 4	Score of 3	Score of 2	Score of 1
Great performance	Good performance	Average performance	Below average	Poor performance

COMPARISON OF VERBAL AND NUMERICAL TEST SCORES FOR 5 CANDIDATES

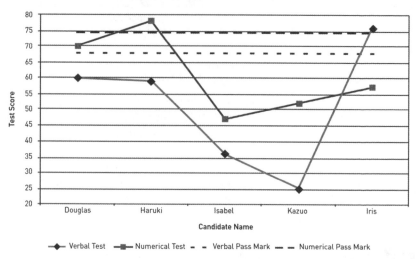

1	Which candidate has the biggest difference between their verbal and numerical test scores?				
A Douglas	B Haruki	C Isabel	D Kazuo	E Iris

2	Estimate the average verbal test score of the 5 candidates.				
A 51	B 61	C 68	D 75	E 255

3	Overall, which candidate performed best on the 2 tests?				
A Douglas	B Haruki	C Isabel	D Kazuo	E Iris

4	By what percentage is Douglas's numerical test score greater than his verbal test score?				
A 6%	B 7%	C 10%	D 14.29%	E 16.67%

5	Approximately how many points would the numerical test pass mark need to be reduced by in order for 60% of this group to pass the test?				
A 15	B 20	C 25	D 56	E 75

COMPARISON OF VERBAL AND NUMERICAL TEST SCORES FOR 5 CANDIDATES

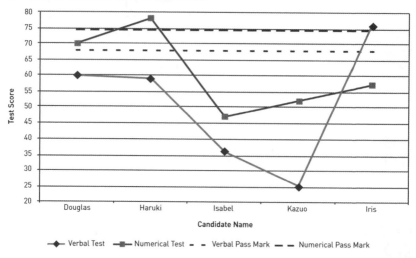

Legend: ◆ Verbal Test ■ Numerical Test - - Verbal Pass Mark — — Numerical Pass Mark

| 1 | Which candidate has the biggest difference between their verbal and numerical test scores? |

Correct answer is D: Kazuo – By looking the graph you can see that the biggest gap between test scores appears to be for Kazuo. If you read off Kazuo's scores from the vertical axis you can see that his verbal test score is equal to 25 and his numerical test score is equal to 52. This gives a gap of 27, which is greater than any of the other candidates.

| 2 | Estimate the average verbal test score of the 5 candidates. |

Correct answer is A: 51 – This question simply requires you to calculate the average score by reading off the values for each candidate from the vertical axis – some scores require a degree of estimation, which is why you're being asked for an estimation of the average. You need to add the 5 verbal test scores together and dividing by their number (5): 60 + 59 + 36 + 25 + 76 = 256. 256 ÷ 5 = 51. 2

Therefore the average verbal test score for this group of candidates is **51**

| 3 | Overall, which candidate performed best on the 2 tests? |

Correct answer is B: Haruki – There are 3 possible ways of answering this question. You can calculate the average test score for each candidate based on their verbal and numerical scores to identify who performed best on average:

Douglas: 65 Kazuo: 38.5
Haruki: **68.5** Iris: 66.5
Isabel: 41.5

Alternatively you could calculate each candidate's total score by adding together their verbal and numerical results and using this to identify the best performance:

Douglas: 130   Kazuo: 77
Haruki: **137**   Iris: 133
Isabel: 83

Or finally you can simply work out the answer from the chart by looking to see the best overall performance as indicated graphically by the results.

Whichever way you do it the answer is the same: **Haruki** performed best overall on the 2 tests.

4	By what percentage is Douglas's numerical test score greater than his verbal test score?

Correct answer is E: 16.67% – Douglas's numerical test score is equal to 70 and his verbal test score is equal to 60, which means that the numerical score is 10 points better. You need to convert this into a percentage: $10 \times 100 \div 60 = 16.67\%$

Therefore, Douglas's numerical test score is **16.67%** greater than his verbal test score.

5	Approximately how many points would the numerical test pass mark need to be reduced by in order for 60% of this group to pass the test?

Correct answer is B: 20 – 60% of a group of 5 candidates is 3 people, therefore you would need to reduce the pass-mark for the numerical test to a level below the scores of the top 3 candidates.

The top 3 scoring candidates on the numerical test are Haruki, Douglas and Iris, who has the lowest score of 57. The pass mark therefore needs to be 56 or lower, but above 52, which is what Kazuo scored.

This is not quite the full answer though – you are asked by how much the pass mark would need to be reduced, not what the new pass mark would be equal to. The pass mark is currently 75 and you need to reduce it to 56 or so. The multiple-choice answer that is the best fit is 20 – this would reduce the pass mark to 55, a point at which Iris would beat it but the remaining 2 candidates would fall below it.

The correct answer is therefore **20**, because the numerical pass mark would need to be dropped 20 points to enable 60% of the group to beat it.

How did you do?

Score of 5	Score of 4	Score of 3	Score of 2	Score of 1
Great performance	Good performance	Average performance	Below average	Poor performance

SO HOW DID YOU DO?

These practice tests are too short individually to make an accurate measurement of your numerical reasoning ability, but if you put your results together we can provide a clearer view. Add your scores on all the 10 practice tests and compare your total score with the table below to gauge your performance.

Total score	What it means	Advice for further development
50	Excellent performance	It looks like you're pretty much there in terms of developing your personal best score. Don't rest on your laurels though. Work through the next chapter of this book, which contains more challenging numerical critical reasoning tests. Completing these more difficult questions will give you an opportunity to stretch your numerical ability to its maximum level. It will also make you feel more confident when you are next faced with a more straightforward, mid-level, numerical reasoning test. Make sure you read the chapters at the end of the book as well – they will provide you with more valuable preparation for performing well at numerical tests.
40–49	Good performance	This level of performance should see you through most numerical tests at this level; but if you feel that it does not reflect your potential personal best score then you might want to do some more practice. Review the advice in Chapter 2 about test strategies and practice, and then have another go at these practice questions.
30–39	Average performance	Numerical reasoning tests are the most widely used level of numerical assessment so an average score is still a positive result. However, you may benefit from more practice to pass numerical tests at this level more effortlessly. Review the advice in Chapter 2 about test strategies and practice. Also review your wrong answers to try to identify any trends in where you go wrong. It might be that you're making simple errors in your calculations or perhaps being too easily taken in by the multiple-choice distracters. Whatever the case, spend some time reviewing where you went wrong and then attempt the practice questions again.

Total score	What it means	Advice for further development
20–29	Below average performance	These are mid-level tests so you might want to polish your numerical ability by correcting any shortfalls in your numeracy abilities by revisiting your work in Chapter 3. Once you are happy that your performance at those foundation level tests reflects your true potential come back to these numerical reasoning tests. To begin with, do not apply the time limits when you attempt these high-level tests – see if that makes a difference. If time is a factor then you should identify which areas of your test-taking strategy might be letting you down. Spend some time developing your raw numerical ability through solving number puzzles and performing mental arithmetic alongside further practice from this book.
0–19	Low performance	

Remember that these 10 practice tests are pitched at the level of difficulty you are most likely to be tested at in the workplace. Numerical tests at this level are often used alongside tests of other abilities, such as verbal reasoning. It's a good idea to balance your performance across different tests so that you can maximise your overall results. Consider working through other books in the Practise & Pass series to help you develop on all your strengths.

CHAPTER 5

NUMERICAL CRITICAL REASONING

N umerical critical reasoning tests are the most advanced form of numerical test. These tests contain complex and high-level numerical information that is designed to simulate the demands of senior-level jobs.

The questions are designed to assess your ability to draw conclusions and solve complex problems about the numerical information in the test. These problems sometimes require you to interpret data that takes some time to make sense of. The correct answers often require multiple stages of calculation and may occasionally demand estimation or other more advanced elements of numerical ability.

The practice questions in this chapter are equivalent to those in real numerical critical reasoning tests. The only differences from the real tests are that there would be more numerical information and questions and a longer time limit in the real tests.

GETTING THE MOST FROM THESE PRACTICE QUESTIONS

The numerical critical reasoning practice questions are grouped together into 10 short practice tests. Each practice test mirrors the approach of a real test: there is a short extract of numerical information followed by a series of five questions for you to answer about the information you have read.

You're encouraged to complete each practice test one at a time and to spend time at the end of each short test to review your responses against the correct answers. Once you are happy with your work, move on to the next test.

The test instructions for these practice tests follow the same format and approach of instructions in real tests – so to be more prepared for the actual tests, make sure that you familiarise yourself with these.

Completing the practice tests

Ideally you should complete the practice tests in conditions that are as close to the real testing environment and experience as possible.

- ▶ Find a quiet place that is free from any distractions.
- ▶ Read the practice test instructions before beginning the first practice test.
- ▶ Real tests have strict time limits. Simulate this by giving yourself eight minutes per practice test. Start timing from the point you begin reading the numerical information for the test.
- ▶ Mark your answers for each practice question by circling the answer you think is correct.
- ▶ You may use a calculator and rough paper to help with your working (do not use a complex, scientific calculator – only basic pocket calculators are allowed in real tests).
- ▶ Do not turn over to the correct answers until you have completed all the practice questions.

Advice for numerical critical reasoning tests

Numerical critical reasoning tests are the most difficult of all numerical tests. When completing a numerical critical reasoning test, remember these points to perform at your personal best:

- ▶ **If you have jumped straight to these practice tests** without completing the lower level numerical tests and numerical reasoning tests we encourage you to stop and go back. Completing the easier tests in turn will greatly benefit your test-taking strategies and level of practice.

▶ **Read the numerical information carefully** before beginning the practice questions. You do not need to memorise it but you will perform better if you are familiar with the information when attempting the questions.

▶ **Never simply write down your answer without checking it first**. Even if you think you have definitely calculated the correct answer, take a moment to check the other choices to make sure you have not been distracted into selecting an incorrect answer.

▶ **Do not simply guess your answers**. You may think that you have a 1 in 5 chance of getting a question correct by guessing, but there is often a scoring mechanism designed to detect or penalise attempts to guess. If you cannot answer a question then review each of the multiple-choice options to identify the ones that are unlikely to be a correct answer. Then you can focus on the shortlist of possible correct answers to identify the one you feel is the right answer.

▶ If you complete all five questions in a practice test within the eight-minute time limit you should **use any time remaining to check your answers** as you would in a real test. This is good test-taking strategy and you should aim to make it a habit.

Before you begin the practice tests

Read the test instructions carefully before you begin – these instructions have been adapted from real test instructions. So you will benefit from familiarity with the language and information they include at the time you actually need it.

Reviewing your answers

Once you have completed a practice test you can turn over the page to check your responses against the correct answers.

Along with each correct answer we have provided some tips and explanations about the answer.

These practice tests are too short to make an accurate measurement of your numerical critical reasoning ability, that's what real tests are for, but we can provide some guidance about what your result means:

Score of 5	Score of 4	Score of 3	Score of 2	Score of 1
Great performance	Good performance	Average performance	Below average performance	Poor performance

If you have scored less than what you would like your personal best score to be then you should look at the pattern of your responses – do you tend to answer specific types of question incorrectly? With numerical critical reasoning tests errors in calculation can occur because of the complex, multi-step nature of the problem-solving process. Sometimes errors are due to particular types of calculation you're less confident about – such as percentages or ratios. Or for many people it can be down to misinterpretation of the numerical information when answering the question.

Here's some advice for avoiding these common types of error.

Work the options	Remember to examine each of the multiple-choice distracters so you can narrow down the range of answers to just two or three likely candidates.
Check the answer against the question	When you have arrived at an answer remember to re-read the question to make sure you have definitely found the result that the question demands.
Check your working!	Just because your answer matches one of the multiple-choice answers doesn't guarantee it's correct. Many multiple-choice distracters are based on common calculation errors. Make sure you have avoided any obvious errors and completed all the calculation steps required to answer what the question asks.

Once you have reviewed your answers take a short break before attempting the next test.

NUMERICAL
CRITICAL
REASONING
PRACTICE TESTS

PRACTICE TEST INSTRUCTIONS

These practice tests each consist of **five** questions about the numerical information given on the first page of each test. You are required to use this numerical information to answer each of the questions in this test. For each question there are five choices of answer labelled A–E – you need to choose one of them. Then mark your chosen answer by circling the appropriate option: A, B, C, D or E.

You may use a calculator and rough paper to help with your working-out.

▸ Allow yourself an 8-minute time-limit for each of the practice tests.
▸ You should work quickly and accurately.
▸ If you are not sure of an answer, fill in what you think, but do not simply guess your answers.
▸ Check your answers against the correct answers given on the pages that follow each practice test.

Remember these important points:

▶ Complete each practice test in a quiet place that is free from any distractions

▶ Start timing the practice test from the point you begin reading the numerical information

▶ Mark your answers for each practice question by fully colouring in the circle for the answer you think is correct

▶ Remember you can use a pocket calculator and rough paper to help with your answers

▶ Do not turn over to the correct answers until you have completed all the practice questions

Now do the first practice test: start timing yourself and read through the numerical information first. Start answering the questions as soon as you are ready.

UTOPIA ONLINE: WEBSITE TRAFFIC ANALYSIS

Minutes spent browsing the *Utopia* Online website

% OF 15,000 VISITORS TO SITE

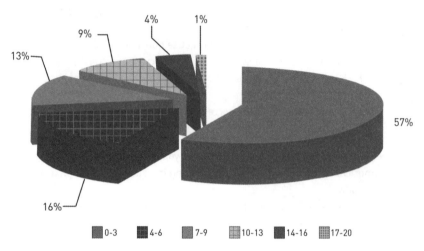

4% 1% 9% 13% 57% 16%

0-3   4-6   7-9   10-13   14-16   17-20

Utopia Online site traffic statistics

Year	Number of visitors	Average number of pages visited
Year 1 (Launch)	50,248	1.25
Year 2	72,811	1.25
Year 3	97,309	1.5
Year 4	127,726	1.75
Year 5	153,436	2
Year 6	182,374	2.25
Year 7 (last year)	210,478	2.5

1	What was the average number of visitors per month in the first 3 years after the launch of the website?			
A 4,187	B 6,121	C 18,364	D 55,092	E 73,456

2	How many of the site visitors spent 3 minutes or less browsing? (Note: give your answer as number of visitors, not a percentage.)			
A 2,400	B 6,450	C 8,000	D 8,550	E 10,950

3	Based on the information given, what would you forecast the average number of pages visited to be in year 10?			
A 2.50	B 2.75	C 3.00	D 3.25	E 3.50

4	What percentage of visitors spent between 4 and 13 minutes browsing the site?			
A 9%	B 13%	C 16%	D 26%	E 38%

5	If the total number of visitors is 210,478, estimate how many browsed the site for less than 10 minutes. (Note: give your answer as number of readers, not the percentage.)			
A 12,900	B 29,467	C 181,011	D 199,954	E 210,478

UTOPIA ONLINE: WEBSITE TRAFFIC ANALYSIS

Minutes spent browsing the *Utopia* Online website

% OF 15,000 VISITORS TO SITE

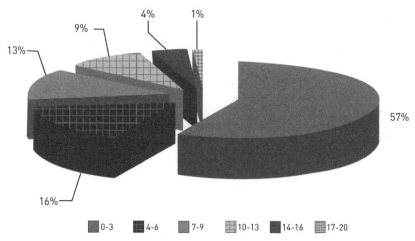

4%
1%
9%
13%
57%
16%

■ 0-3 ■ 4-6 ■ 7-9 ▦ 10-13 ■ 14-16 ▦ 17-20

Utopia Online site traffic statistics

Year	Number of visitors	Average number of pages visited
Year 1 (Launch)	50,248	1.25
Year 2	72,811	1.25
Year 3	97,309	1.5
Year 4	127,726	1.75
Year 5	153,436	2
Year 6	182,374	2.25
Year 7 (last year)	210,478	2.5

1 What was the average number of visitors per month in the first 3 years after the launch of the website?

Correct answer is B: 6,121 – You need to add the number of visitors from years 1–3 and then divide by the number of years to get the average:
50,248 + 72,811 + 97,309 = 220,368 (total visitors for years 1–3)
220,368 ÷ 3 = 73,456 (average visitors per year)
73,456 ÷ 12 = **6,121** (average visitors per month for years 1–3)

2 How many of the site visitors spent 3 minutes or less browsing? (Note: give your answer as number of visitors, not the percentage.)

Correct answer is D: 8,550 – You can see from the pie chart that 57% of visitors spent 3 minutes or less browsing the site. You can also see from the title of the pie chart that 15,000 visitors were included in the survey. To answer the question you must therefore calculate what 57% of 15,000 people is equal to: 57 × 15,000 = 855,000. 855,000 ÷ 100 = **8,550** visitors

This question was fairly kind to you because it gave explicit instructions not to give the answer as a percentage, which would have been an easy mistake to make.

3 Based on the information given, what would you forecast the average number of pages visited to be in year 10?

Correct answer is D: 3.25 – This question draws on your sequential reasoning skills to examine the rate of increase in page visits each year. If you compare each year's number you can see that since year 2 the average number of pages visited has increased by 0.25 each year. If this increase per year continues the averages for years 8, 9 and 10 will be as follows:
Year 8: 2.50 + 0.25 = 2.75
Year 9: 2.75 + 0.25 = 3.00
Year 10: 3.00 + 0.25 = **3.25**

4 What percentage of visitors spent between 4 and 13 minutes browsing the site?

Correct answer is E: 38% – The first step is to correctly interpret the pie chart and extract the appropriate values; you need to add the values from the categories 4 – 6, 7 – 9 and 10 – 13: 16% + 13% + 9% = **38%** of visitors

5 If the total number of visitors is 210,478, estimate how many browsed the site for less than 10 minutes. (Note: give your answer as number of readers, not the percentage.)

Correct answer is C: 181,011 – This question introduces some additional information to the values in the pie chart and table; in this case the total number of visitors. To answer the question you need to apply the appropriate percentages from the pie chart to this new total number of visitors. First you need to calculate what percentage of visitors spent less than 10 minutes browsing. This means adding the values for the categories 0 – 3, 4 – 6, 7 – 9: 57% + 16% + 13% = 86%

Because the question requires this answer as a number and not a percentage you need to calculate what 86% of 210,478 is equal to: 86 ÷ 100 = 0.86. 0.86 × 210,478 = **181,011** visitors

How did you do?

Score of 5	Score of 4	Score of 3	Score of 2	Score of 1
Great performance	Good performance	Average performance	Below average	Poor performance

UTOPIA MAGAZINE: AVERAGE MONTHLY CIRCULATION DURING GIVEN YEAR

		Cover price
Year 1 (launch)	50,248	£1.25
Year 2	72,811	£1.25
Year 3	97,309	£1.50
Year 4	127,726	£1.75
Year 5	153,436	£2.00
Year 6	182,374	£2.25
Year 7 (last year)	210,478	£2.50

Utopia magazine readership survey: Reading habits

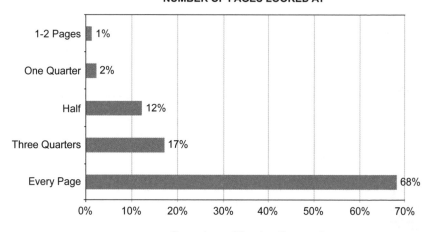

NUMBER OF PAGES LOOKED AT

- 1-2 Pages 1%
- One Quarter 2%
- Half 12%
- Three Quarters 17%
- Every Page 68%

0% 10% 20% 30% 40% 50% 60% 70%

Percentage of Readers Surveyed

1	What was the percentage increase in average monthly circulation from year 1 to year 2?			
A	B	C	D	E
19.61%	30.99%	93.66%	44.90%	44%

2	Assuming the whole circulation paid the full cover price, what was the average monthly income from *Utopia* sales during year 5 to year 7?			
A	B	C	D	E
£526,195	£136,572	£455,240	£409,716	£414,469

3	Based on the information given, what would you forecast the cover price to be in year 10?			
A	B	C	D	E
£3.25	£2.75	£3.00	£2.50	£3.50

4	If the total readership is 210,478, estimate how many look at 75% or more of the pages. (Note: Give your answer as number of readers, not the percentage.)			
A	B	C	D	E
12,750	178,906	157,859	35,781	31,572

5	Express as a ratio the relationship of readers who look at half the magazine and readers who look at one-quarter (half:quarter).			
A	B	C	D	E
6:1	7:2	2:1	4:1	5:1

UTOPIA MAGAZINE: AVERAGE MONTHLY CIRCULATION DURING GIVEN YEAR

		Cover price
Year 1 (launch)	50,248	£1.25
Year 2	72,811	£1.25
Year 3	97,309	£1.50
Year 4	127,726	£1.75
Year 5	153,436	£2.00
Year 6	182,374	£2.25
Year 7 (last year)	210,478	£2.50

Utopia magazine readership survey: Reading habits

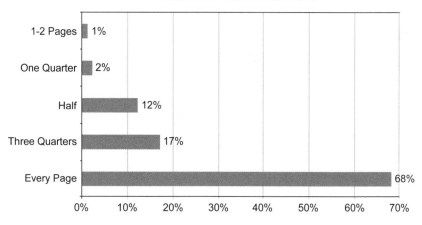

NUMBER OF PAGES LOOKED AT

- 1-2 Pages: 1%
- One Quarter: 2%
- Half: 12%
- Three Quarters: 17%
- Every Page: 68%

Percentage of Readers Surveyed

---

1 | What was the percentage increase in average monthly circulation from year 1 to year 2?

Correct answer is D: 44.90% – The first step is to calculate the difference between the circulation figures for the 2 years: Year 1: 50,248, Year 2: 72,811: 72,811 – 50,248 = 22,563.

Next, you need to convert this to a percentage: 22,563 ÷ 50,248 × 100 = 44.90

Therefore, the percentage increase in average monthly circulation from year 1 to year 2 is **44.90%**.

2	Assuming the whole circulation paid the full cover price, what was the average monthly income from *Utopia* sales during year 5 to year 7?

Correct answer is E: £414,469 – The first step requires you to calculate 3 sub-totals – the income from each year from year 5 to year 7 based on the circulation and cover price for each year:

Year 5: 153,436 × £2.00 = £306,872
Year 6: 182,374 × £2.25 = £410,341.50
Year 7: 210,478 × £2.50 = £526,195

Now you must add these 3 monthly income figures together and divide by 3 to get the average: £306,872 + £410,341.50 + £526,195 = £1,243,408. £1,243,408 ÷ 3 = £414,469.33

Therefore, the average monthly income from *Utopia* sales during year 5 to year 7 is equal to **£414,469**

3	Based on the information given, what would you forecast the cover price to be in year 10?

Correct answer is A: £3.25 – By the look of years 2–7 the cover price has increased in regular increments of 25p per year. So we can predict the year 10 cover price by applying this trend to years 8, 9 and 10:

Year 7: £2.50   Year 9: £3.00
Year 8: £2.75   Year 10: £3.25

We would therefore forecast the cover price in year 10 to be **£3.25**

4	If the total readership is 210,478, estimate how many look at 75% or more of the pages. (Note: Give your answer as a number of readers, not a percentage.)

Correct answer is B: 178,906 – This question requires you to use the chart rather than the table of numerical information to find the values you need. You can see from the chart that the percentage of readers who read more than 75% of the magazine fall into 2 groups. You need to add these percentages: 3/4 of magazine + every page = 17% + 68% = 85%

We have also been told that the total readership is equal to 210,478; so we need to calculate what 85% of this total is equal to: 85 × 210,478 ÷ 100 = 178,906.3

Therefore 178,906 readers look at 75% or more of the pages.

5	Express as a ratio the relationship of readers who look at half the magazine and readers who look at one-quarter (half:quarter).

Correct answer is A: 6:1 – The starting point is to establish the percentages in each of these 2 groups: Half the magazine = 12%, One-quarter of the magazine = 2%

The ratio of half:quarter is therefore 12:2, which should be expressed as 6:1

How did you do?

Score of 5	Score of 4	Score of 3	Score of 2	Score of 1
Great performance	Good performance	Average performance	Below average	Poor performance

UTOPIA MAGAZINE READERSHIP SURVEY: TIME SPENT READING

Readers surveyed = 15,000

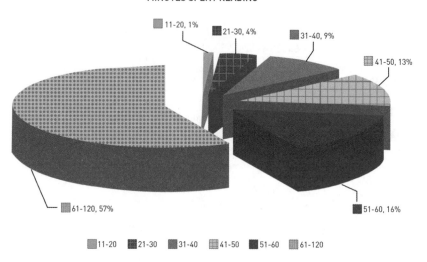

MINUTES SPENT READING

11-20, 1%   21-30, 4%   31-40, 9%   41-50, 13%

61-120, 57%   51-60, 16%

■ 11-20   ■ 21-30   ■ 31-40   ■ 41-50   ■ 51-60   ■ 61-120

1	What percentage of readers surveyed spent between 31 and 60 minutes reading the magazine?

A	B	C	D	E
62%	38%	9%	13%	16%

2	How many of the readers surveyed spent more than an hour reading?

A	B	C	D	E
10,950	6,450	8,550	2,400	8,000

3	How many of the readers surveyed spent half an hour or less reading?

A	B	C	D	E
2,100	150	750	6,450	700

4	If the total readership is 210,478, estimate how many read the magazine for over 40 minutes.

A	B	C	D	E
29,467	18,943	12,900	181,011	199,954

5	If reading the whole magazine takes on average 45 minutes, using the survey sample, estimate how many people finish reading the magazine each month.

A	B	C	D	E
176	11,925	14,250	15,000	169,435

UTOPIA MAGAZINE READERSHIP SURVEY: TIME SPENT READING

Readers surveyed = 15,000

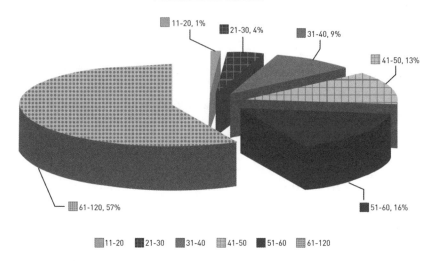

MINUTES SPENT READING

11-20, 1%   21-30, 4%   31-40, 9%   41-50, 13%

61-120, 57%   51-60, 16%

11-20   21-30   31-40   41-50   51-60   61-120

---

1   **What percentage of readers surveyed spent between 31 and 60 minutes reading the magazine?**

Correct answer is B: 38% – This is a basic warm-up question to help you to become familiar with the pie chart. You need to add the percentages from the groups within this range of 31–60 minutes:
31–40 minutes = 9%
41–50 minutes = 13%
51–60 minutes = 16%
9 + 13 + 16 = **38**

Therefore, **38%** of readers surveyed spent between 31 and 60 minutes reading the magazine.

---

2   **How many of the readers surveyed spent more than an hour reading?**

Correct answer is C: 8,550 – You can see from the pie chart that 57% of the readers surveyed spent over an hour reading (the 61–120 minutes group). You therefore need to use this percentage along with the number of readers surveyed (15,000) to calculate the number of readers as follows: 57 × 15,000 ÷ 100 = 8,550

Therefore **8,550** readers surveyed spent more than an hour reading.

3	How many of the readers surveyed spent half an hour or less reading?

Correct answer is C: 750 – You can see from the pie chart that 5% of the readers surveyed spent half an hour or less reading (the 11–30 minutes groups). You therefore need to use this percentage along with the number of readers surveyed (15,000) to calculate the number of readers (note that none of the multiple-choice options are percentages): 5 × 15,000 ÷ 100 = 750

Therefore **750** readers surveyed spent half an hour or less reading.

4	If the total readership is 210,478, estimate how many read the magazine for over 40 minutes.

Correct answer is D: 181,011 – Use the 41–50 minutes group as the start point. You can see from the pie chart that 86% of the readers surveyed spent over 40 minutes reading (the 41–120 minutes groups). You therefore need to use this percentage along with the number for the total readership (210,478 not 15,000) to calculate the number of readers: 86 × 210,478 ÷ 100 =181,011.08

Therefore **181,011** readers are likely to have spent more than 40 minutes reading the magazine based on a total readership of 210,478

5	If reading the whole magazine takes on average 45 minutes, using the survey sample, estimate how many people finish reading the magazine each month.

Correct answer is B: 11,925 – This question is tricky because the 45 minutes figure falls in the middle of one of the groups in the pie chart. Your best bet is to take the 41–50 minutes group and divide its percentage by 2 as a way of estimating how many in this group spend more than 45 minutes, which gives you 6.5%. You then need to add this percentage to the remaining percentages in the groups that spent longer than this. 6.5 + 16 + 57 = 79.5%

You then need to use this percentage along with the number of readers surveyed (15,000) to calculate the number of readers: 79.5 × 15,000 ÷ 100 = 11,925

Therefore, approximately **11,925** readers surveyed read the whole magazine.

How did you do?

Score of 5	Score of 4	Score of 3	Score of 2	Score of 1
Great performance	Good performance	Average performance	Below average	Poor performance

Book sales by category

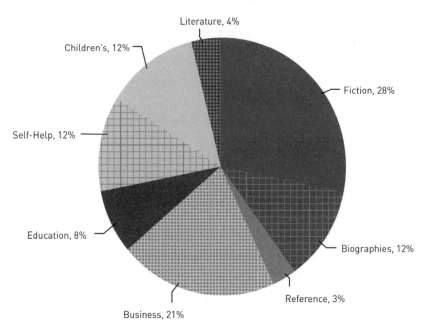

BOOK SALES

Literature, 4%

Children's, 12%

Fiction, 28%

Self-Help, 12%

Education, 8%

Biographies, 12%

Reference, 3%

Business, 21%

1	Express the balance of Literature to Biography sales as a ratio.

A	B	C	D	E
1:2	1:3	3:4	4:6	Cannot say

2	If Reference is worth £18 million this year, on average how much is it worth per week?

A	B	C	D	E
£346,154	£350,000	£375,000	£1,500,000	Cannot say

3	If the total revenue for Education is £48 million, what is the total revenue for Business worth?

A	B	C	D	E
£48 million	£81 million	£126 million	£229 million	Cannot say

4	If the total revenue from Fiction in month 1 was £1,400,000, what was the average cost per item?

A	B	C	D	E
£7.99	£15	£26.92	£28	Cannot say

5	It is estimated that Children's will account for 15% of sales next year and 20% the year after. If Children's is worth £90 million this year, and total sales are predicted to increase by 5% each year, what will be its estimated worth in 2 years' time?

A	B	C	D	E
£108 million	£165 million	£750 million	£825 million	Cannot say

Book sales by category

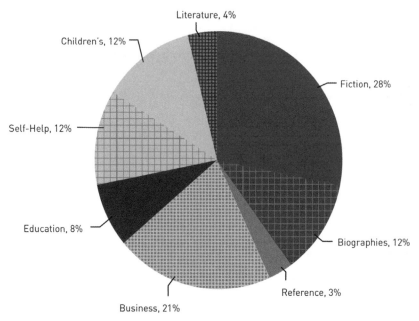

BOOK SALES

Literature, 4%

Children's, 12%

Fiction, 28%

Self-Help, 12%

Education, 8%

Biographies, 12%

Reference, 3%

Business, 21%

1	Express the balance of Literature to Biography sales as a ratio.

Correct answer is B: 1:3 – Literature accounts for 4% of book sales while Biography accounts for 12%. The ratio is therefore 4:12, which should be expressed as **1:3**

2	If Reference is worth £18 million this year, on average how much is it worth per week?

Correct answer is A: £346,154 – You need to divide the annual figure (18 million) by 52 to arrive at an average weekly value: £18,000,000 ÷ 52 = £346,153.84

Therefore on average Reference book sales are worth **£346,154** per week

3

If the total revenue for Education is £48 million, what is the total revenue for Business worth?

Correct answer is C: £126 million – Education accounts for 8% of sales. If we know that this amounts to £48 million we can calculate the value of 1% and therefore the 21% that Business sales are worth: £48,000,000 ÷ 8 = £6,000,000. £6,000,000 × 21 = £126,000,000

Therefore, Business sales are worth **£126 million**

4

If the total revenue from Fiction in Month 1 was £1,400,000, what was the average cost per item?

Correct answer is E: Cannot say – There is no way of calculating this answer based on the information given. If the Cannot say distracter is presented on every question in a test there is a reasonably strong likelihood that you will need to use it at some point to answer a question correctly.

5

It is estimated that Children's will be worth 15% next year and 20% the year after. If Children's is worth £90 million this year, and total sales are predicted to increase by 5% each year, what will be its estimated worth in 2 years' time?

Correct answer is B: £165 million – The first step is to unpack the information in the question. We are being asked to look ahead 2 years. The position at that time will be as follows:
Children's = 20% of total book sales
Total book sales – increase by 10%

We can estimate this year's total book sales by using the Children's £90 million figure and its share of total book sales (12%): Total book sales this year = £90,000,000 × 100 ÷ 12 = £750,000,000

In 2 years this figure will have increased by 10%, which will bring total book sales to £825 million

Meanwhile, Children's will have increased to 20% of the total book sales (£825 million): £825,000,000 ÷ 100 × 20 = £165,000,000

Therefore the estimated worth of the children's category of book sales will be **£165 million** in 2 years' time.

How did you do?

Score of 5	Score of 4	Score of 3	Score of 2	Score of 1
Great performance	Good performance	Average performance	Below average	Poor performance

Book sales by category

BOOK SALES

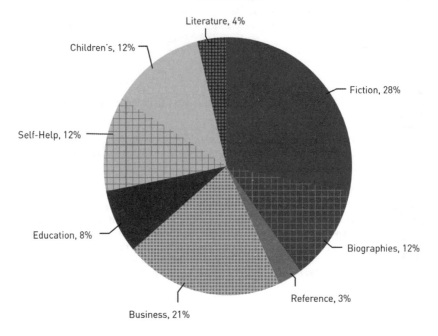

1 Express as a fraction the proportion of total book sales that Children's and Education account for when combined.

A	B	C	D	E
½	²/₃	³/₄	⅕	Cannot say

2 If sales of electronic books decrease overall book sales by 12% next year, what percentage of book sales will be accounted for by Fiction?

A	B	C	D	E
2%	4%	8%	16%	Cannot say

3 Revenue from Children's for this year is £72 million, which is an increase of 8.75% from last year. What was Children's worth last year?

A	B	C	D	E
£6 million	£63 million	£66 million	£85 million	Cannot say

4 If the total revenue from book sales increases by ¼ next year how much will Self-help be worth if Self-help sales are worth £108 million this year and they continue to account for 18% of sales?

A	B	C	D	E
£111 million	£119 million	£127 million	£135 million	Cannot say

5 Fiction sales are worth an average of £42 million per quarter of this year. What would Fiction sales have been for the first three quarters of last year if sales have increased by 13% this year from last year's figures?

A	B	C	D	E
£37 million	£84 million	£112 million	£126 million	Cannot say

Book sales by category

BOOK SALES

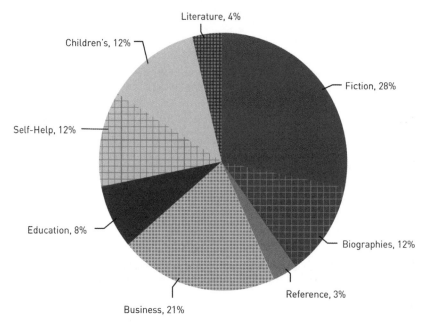

Literature, 4%

Children's, 12%

Fiction, 28%

Self-Help, 12%

Education, 8%

Biographies, 12%

Reference, 3%

Business, 21%

---

1 Express as a fraction the proportion of total book sales that Children's and Education account for when combined.

Correct answer is D ⅕ – The combined percentage of Children's and Education contribution to book sales is 12% + 8% = 20%

To express 20% as a fraction you can follow these steps: 20% is the same as the fraction $^{20}/_{100}$. The fraction $^{20}/_{100}$ can be reduced in size by dividing the right side by the left side: 100 divided by 20 = $^{1}/_{5}$

Therefore the correct answer is ⅕

---

2 If sales of electronic books decrease overall book sales by 12% next year, what percentage of book sales will be accounted for by Fiction?

Correct answer is E: Cannot say – There is no way of calculating this answer based on the information given.

3 Revenue from Children's for this year is £72 million, which is an increase of 8.75% from last year. What was Children's worth last year?

Correct answer is C: £66 million – Another way of looking at this numerical information is that revenue from Children's is worth 108.75% of last year's sales. So we can use this figure along with the value in pounds to calculate what 100% is equal to: £72 ÷ 108.75 = 0.66
0.66 × 100 = £66

Therefore, Children's sales were worth **£66 million** last year – 8.5% less than this year.

4 If the total revenue from book sales increases by ¼ next year how much will Self-help be worth if Self-help sales are worth £108 million this year and they continue to account for 12% of sales?

Correct answer is D: £135 million – If total book sales next year increase by ¼ (25%) then we can apply the same increase to Self-help books because we are told that they will still account for an 12% share of this increased total.

25% of £108,000,000: £108,000,000 ÷ 4 = £27,000,000
£108,000,000 + £27,000,000 = £135,000,000

Therefore next year's sales of self-help books is likely to be worth **£135 million** if total book sales increase by ¼

5 Fiction sales are worth an average of £42 million per quarter of this year. What would Fiction sales have been for the first three quarters of last year if sales have increased by 13% this year from last year's figures?

Correct answer is C: £112 million – First we need to work out what the average sales of Fiction would have been worth per quarter of last year. We know that they are worth £42 million per quarter this year and that this is a 13% increase on the previous year. Another way of stating this position is that at £42 million per quarter, this year's Fiction sales represent 113% of last year's figures. We can use this to calculate what 100% would be, which is the equivalent of the previous year's figure per quarter:

£42 ÷ 113 = 0.37
0.37 × 100 = £37

Therefore last year Fiction sales were worth £37 million per quarter.

But we're not quite there yet; the question asked us to calculate the value of fiction sales for quarters 1 to 3 of last year. We therefore need to multiply this figure by 3: £37,000,000 × 3 = £111.5,000,000

Therefore the worth of the Fiction category of book sales in the first 3 quarters of last year is **£112 million**

How did you do?

Score of 5	Score of 4	Score of 3	Score of 2	Score of 1
Great performance	Good performance	Average performance	Below average	Poor performance

UTOPIA MAGAZINE: BUSINESS PERFORMANCE VS LAST YEAR'S RESULTS

	£000s	
	2009	**2010**
Sales	£2,498	£3,347
Cost of goods	£2,102	£2,835
Gross profit	£396	£512
Production costs		
Wages	£197	£268
Distribution	£78	£86
Utilities	£19	£21
Repairs and maintenance	£12	£11
Head office	£39	£48
Net profit	£51	£78

1 What was the average increase in production costs between 2009 and 2010?

A	B	C	D	E
£3,400	£17,800	£33,555	£89,000	£155,800

2 By what percentage has net profit increased between 2009 and 2010?

A	B	C	D	E
27%	44%	47%	53%	54%

3 If the Head office costs increase by 4.5% next year, what would these costs be in 2011?

A	B	C	D	E
£40,755	£45,840	£48,450	£50,160	£58,666

4 If the Cost of goods increased by the same amount each year, how much was the Cost of goods in 2008?

A	B	C	D	E
£1,369,000	£1,649,000	£1,764,000	£2,101,000	£2,835,000

5 If Distribution costs increase by 5% next year, what percentage of Gross profits will the Net profits be worth?

A	B	C	D	E
4.3%	7%	14%	20%	24%

UTOPIA MAGAZINE: BUSINESS PERFORMANCE VS LAST YEAR'S RESULTS

	£000s	
	2009	**2010**
Sales	£2,498	£3,347
Cost of goods	£2,102	£2,835
Gross profit	£396	£512
Production costs		
Wages	£197	£268
Distribution	£78	£86
Utilities	£19	£21
Repairs and maintenance	£12	£11
Head office	£39	£48
Net profit	£51	£78

1	What was the average increase in production costs between 2009 and 2010?

Correct answer is B: £17,800 – The first step is to work out the average production costs for 2009 and 2010; this requires you to add the individual costs together and divide their sum by 5 (their number):
2009: £197 + £78 + £19 + £12 + £39 = £345 ÷ 5 = £69
2010: £268 + £86 + £21 + £11 + £48 = £434 ÷ 5 = £86.8

You now need to subtract one sub-total from the other: £86.8 – £69 = £17.8

The figures in the table are in thousands, therefore the correct answer is **£17,800**

2	By what percentage has Net profit increased between 2009 and 2010?

Correct answer is D: 53% – You can work out the percentage by first calculating the increase in pounds: £78 – £51 = £27

Therefore Net profit has increased by £27,000 between 2009 and 2010. You can use this information to calculate the increase in percentage terms: £27 × 100 ÷ 51 = £52.94

Therefore Net profit increased by **53%** in 2010

| **3** | If the Head office costs increase by 4.5% next year, what would these costs be in 2011? |

Correct answer is D: £50,160 – Head office costs in 2010 were £48,000, so to calculate the 2011 value based on an increase of 4.5% you first work out the value of 4.5% in pounds: 4.5 × 48 ÷ 100 = 2.16

You then add this to the 2010 figure: £48,000 + £2,160 = £50,160

Therefore, Head office costs in 2011 will be **£50,160** based on an increase of 4.5% on the 2010 figure.

| **4** | If the Cost of goods increased by the same amount each year, how much was the Cost of goods in 2008? |

Correct answer is A: £1,369,000 – To answer this one you need to subtract the 2009 figure from the 2010 figure and then subtract this result from the 2009 value to arrive at the 2008 number: 2010 – 2009: £2,835 – £2,102 = £733

Because the Cost of goods has increased by the same value each year we can apply this to the 2009 figure to get the 2008 value: 2008 Cost of goods = £2102 – £733 = £1,369

Therefore the 2008 Cost of goods value is **£1,369,000**

| **5** | If Distribution costs increase by 5% next year, what percentage of Gross profits will the Net profits be worth? |

Correct answer is C: 14% – The first step is to calculate the increased distribution costs based on the 5% rise for next year: Distribution cost in 2011: 86 ÷ 100 × 5 = 4.3

We can apply this increased cost to the 2010 Net profit: 78 – 4.3 = 73.7. So if only distribution costs change next year the Net profit will be £73,700

We next need to calculate what percentage this Net profit is of the Gross profit: 73.7 ÷ 512 × 100 = 14.39%

Therefore, if distribution costs increase by 5% next year, the magazine's Net profit will be worth **14%** of its Gross profit.

How did you do?

Score of 5	Score of 4	Score of 3	Score of 2	Score of 1
Great performance	Good performance	Average performance	Below average	Poor performance

UTOPIA MAGAZINE: BUSINESS PERFORMANCE VS LAST YEAR'S RESULTS

	£000s	
	2009	2010
Sales	£2,498	£3,347
Cost of goods	£2,102	£2,835
Gross profit	£396	£512
Production costs		
Wages	£197	£268
Distribution	£78	£86
Utilities	£19	£21
Repairs and maintenance	£12	£11
Head office	£39	£48
Net profit	£51	£78

1	What percentage of the Gross profit was spent on Wages and Distribution in 2009?				
A	B	C	D	E
47.45%	53.71%	69.14%	69.44%	85.06%

2	If the Gross profit continues to increase by the same amount each year, what would it be by May 2011?				
A	B	C	D	E
£105,000	£396,000	£512,000	£628,000	£908,000

3	If in 2010 the Cost of goods increased by 3%, Sales increased by 3.9% and Production costs increased by 1.5%, what would be the Net profit in 2010?				
A	B	C	D	E
£39,030	£116,970	£169,560	£208,530	£300,090

4	Which production cost saw the biggest percentage rise between 2009 and 2010?				
A	B	C	D	E
Wages	Distribution	Utilities	Repairs and maintenance	Head office

5	Express the 2009 production costs as a percentage of the 2009 Gross Profit.				
A	B	C	D	E
0.51%	12.88%	77%	87.12%	345%

UTOPIA MAGAZINE: BUSINESS PERFORMANCE VS LAST YEAR'S RESULTS

	£000s	
	2009	**2010**
Sales	£2,498	£3,347
Cost of goods	£2,102	£2,835
Gross profit	£396	£512
Production costs		
Wages	£197	£268
Distribution	£78	£86
Utilities	£19	£21
Repairs and maintenance	£12	£11
Head office	£39	£48
Net profit	£51	£78

---

1   **What percentage of the Gross profit was spent on Wages and Distribution in 2009?**

Correct answer is D: 69.44% – The first step is to add the 2009 costs for Wages and Distribution: 197 + 78 = 275

Next we need to convert this into a percentage of the 2009 Gross profit (396): 275 ÷ 396 × 100 = 69.44

Therefore, **69.44%** of the Gross profit was spent on Wages and Distribution in 2009.

---

2   **If the Gross profit continues to increase by the same amount each year, what would it be by May 2011?**

Correct answer is D: £628,000 – To answer this one you need to subtract the 2009 Gross profit figure from the 2010 figure, and then add the result to the 2010 value to arrive at the 2011 number: 2010 – 2009: £512 – £396 = £116

Because Gross profit has increased by the same value each year we can apply this to the 2010 figure to get the 2011 value: 2011 Gross profit = £512 + £116 = £628

Therefore the 2011 Gross profit value is **£628,000**

<table>
<tr><td>**3**</td><td>If in 2010 the Cost of goods increased by 3%, Sales increased by 3.9% and Production costs increased by 1.5%, what would be the Net profit in 2010?</td></tr>
</table>

Correct answer is B: £116,970 – First we need to calculate the new costs based on these percentage increases:
Cost of goods 3% increase: $3 \times £2,835 \div 100 = £85.05$
Production costs 1.5% increase: $1.5 \times £434 \div 100 = £6.51$

So the total increase in costs is £85.05 + £6.51 = £91.56

There has also been an increase in Sales of 3.9%: $3.9 \times £3,347 \div 100 = £130.53$
So we now need to apply these changes to the Net profit figure: The costs must be subtracted from Net profit: £78 – £91.56 = –£13.56
The sales increase must be added to the new Net profit: –£13.56 + £130.53 = £116.97

Therefore the revised 2010 Net profit figure is **£116,970**

<table>
<tr><td>**4**</td><td>Which production cost saw the biggest percentage rise between 2009 and 2010?</td></tr>
</table>

Correct answer is A: Wages – You might be able to exclude some obvious wrong answers using mental arithmetic by rejecting any costs that obviously do not have the biggest increase. Or you can calculate the percentage increase for each cost by first working out the increases in pounds:
Wages: £268 – £197 = £71
Distribution: £86 – £78 = £8
Utilities: £21 – £19 = £2
Repairs and maintenance: £11 – £12 = –£1
Head Office: £48 – £39 = £9

At this stage you're probably safe to exclude Repairs and maintenance from your answers as the costs decreased from 2009 to 2010 and it is therefore unlikely to be the biggest increase of the costs!

Next you need to convert these increases into percentages for the remaining costs:
Wages: £71 ÷ £197 $\times$ 100 = **36%**
Distribution: £8 ÷ £78 $\times$ 100 = 10%
Utilities: £2 ÷ £19 $\times$ 100 = 11%
Head Office: £9 ÷ £39 $\times$ 100 = 23%

Therefore the biggest increase was in Wages at 36% from 2009 to 2010. This company obviously needs a pay freeze in 2011.

<table>
<tr><td>**5**</td><td>Express the 2009 production costs as a percentage of the 2009 Gross Profit.</td></tr>
</table>

Correct answer is D: 87.12% – The first step is to calculate the total of the 2009 production costs. You can do this by adding together all the production cost values in the table. Or a quicker way of doing it is to simply subtract net profit from the gross profit:

Total production costs = Gross Profit (396) – Net Profit (51) = 345

Next you need to convert this into a percentage using the Gross Profit value:
345 ÷ 396 $\times$ 100 = 87.12%

Therefore, the 2009 production costs are equal to **87.12%** of the Gross Profit.

How did you do?

Score of 5	Score of 4	Score of 3	Score of 2	Score of 1
Great performance	Good performance	Average performance	Below average	Poor performance

Sales (£000s) 2009 vs 2010

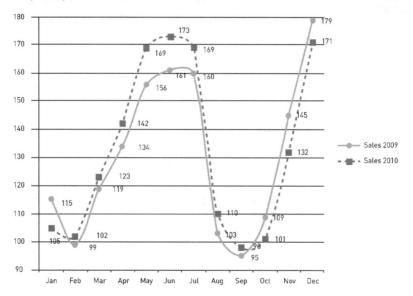

1	What was the difference in total sales between 2009 and 2010?			
A £20,000	B £200,000	C £1,575,000	D £1,595,000	E £2,000,000

2	By what percentage did sales change between January 2009 and January 2010?			
A 8.7% increase	B 8.7% decrease	C 9.5% increase	D 9.5% decrease	E No change

3	Between which two months in 2010 was there the greatest increase in sales?			
A February and March	B March and April	C July and August	D September and October	E November and December

4	In 2009 Valentine's promotion lines accounted for 9% of January's sales and 21.5% of February's sales. During those two months how much of the sales results were not due to the Valentine's promotion lines?			
A £83,360	B £182,365	C £203,060	D £214,000	E £245,640

5	A manager in the business received a bonus of £85 for every £100,000 in total yearly sales. How much more in bonuses did the manager receive in 2010 compared with 2009?			
A £20	B £100	C £200	D £1000	E £20,000

Sales (£000s) 2009 vs 2010

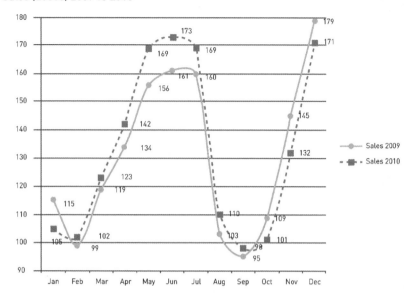

---

1   What was the difference in total sales between 2009 and 2010?

Correct answer is A: £20,000 – The first step is to add together the 2009 and 2010 sales for each month to get the total sales:

2009 total sales: £115 + £99 + £119 + £134 + £156 + £161 + £160 + £103 + £95 + £109 + £145 + £179 = £1,575

2010 total sales: £105 + £102 + £123 + £142 + £169 + £173 + £169 + £110 + £98 + £101 + £132 + £171 = £1,595

You can now subtract the 2009 figure from the 2010 figure: £1,595 – £1,575 = £20. Because the chart is in thousands this result is equal to £20,000

Therefore, the difference in total sales between 2009 and 2010 is **£20,000**

---

2   By what percentage did sales change between January 2009 and January 2010?

Correct answer is D: 9.5% decrease – To answer this one you need to subtract the January 2010 sales figure from the January 2009 figure and then convert it into a percentage: Jan 2009 sales – Jan 2010 sales: £115 – £105 = £10. In percentage terms this works out as: £10 ÷ £105 × 100 = 9.52%

The 2009 figure was higher than 2010, therefore there was a **9.5%** decrease in sales between Jan 2009 and Jan 2010.

3	Between which two months in 2010 was there the greatest increase in sales?

Correct answer is E: November and December – You can reduce the quantity of calculation you need to do by only calculating the increases for the months presented in the multiple-choice options:
February – March: £123 – £102 = £21
March – April: £142 – £123 = £19
July – August: You can see from the graph that sales decreased in August
September – October: £101 – £98 = £3
November – December: £171 – £132 = **£39**

Therefore the greatest increase in sales in 2010 was £39,000 between **November and December**.

4	In 2009 Valentine's promotion lines accounted for 9% of January's sales and 21.5% of February's sales. During those two months how much of the sales results were not due to the Valentine's promotion lines?

Correct answer is B: £182,365 – First we need to subtract non-Valentine's sales from the January and February 2009 figures:
January 2009 sales = £115
February 2009 sales = £99
Total sales for the 2 months = £214

9% of Jan 2009 sales: £115 ÷ 100 × 9 = £10.35
21.5% of Feb 2009 sales: £99 ÷ 100 × 21.5 = £21.29
Total Valentine's sales for the 2 months: £10.35 + £21.29 = £31.64

Therefore the sales in January and February 2009 that were not related to Valentine's sales were £214 – £31.64 = £182.36 or **£182,365**

5	A manager in the business received a bonus of £85 for every £100,000 in total yearly sales. How much more in bonuses did the manager receive in 2010 compared with 2009?

Correct answer is A: £20 – Luckily we calculated the sales totals for 2009 and 2010 for Question 1 – so if you were using good test-taking strategy you would have made a note of these at the time. If you didn't then you need to calculate them again:
2009 total sales: £115 + £99 + £119 + £134 + £156 + £161 + £160 + £103 + £95 + £109 + £145 + £179 = £1,575
2010 total sales: £105 + £102 + £123 + £142 + £169 + £173 + £169 + £110 + £98 + £101 + £132 + £171 = £1,595

To calculate the sales bonus for each year you need to divide these figures by 100,000 (remembering to factor each up by 100,000 first, because the graph is in thousands):
2009 sales bonus: £1,575,000 ÷ 100,000 = £1,575
2010 sales bonus: £1,595,000 ÷ 100,000 = £1,595

When you subtract one bonus figure from the other you see that the manager earned £20 more in 2010 than in 2009. It probably didn't change the manager's life too much.

How did you do?

Score of 5	Score of 4	Score of 3	Score of 2	Score of 1
Great performance	Good performance	Average performance	Below average	Poor performance

COMPARABLE MAGAZINE CIRCULATION FIGURES

For the last 12-month period

Magazine name	Average monthly circulation during given year	UK cover price
Utopia	210,478	£2.50
Green Earth	166,432	£1.80
New Age	1,129,407	£2.80
Natural Living	104,003	£1.15
Networker	78,024	£1.10
Sign of the Times	48,075	£4.20
One People	47,036	£3.15
Plan-it Power	38,143	£2.00

$$\text{Production cost per unit (£s)} = \frac{100,000}{\text{Average monthly circulation} + 20,000}$$

1 What is the price of *Utopia* expressed as a percentage of the price of the most expensive magazine?

A	B	C	D	E
59.52%	79.37%	89.29%	100%	126%

2 Assuming the whole circulation paid the full cover price, what was the average monthly income from sales during the last 12-month period for the three magazines with the highest circulations?

A	B	C	D	E
£299,577.60	£332,342.68	£526,1958	£1,329,370.73	£3,988,112.20

3 If 5% of the cover price is paid to newsagents for each *Utopia* magazine sold, and *Utopia* is a monthly magazine, estimate from the given information how much was paid to newsagents in total over the last 12 months.

A	B	C	D	E
£1.50	£2.50	£26,309.75	£315,717	£526,195

4 Which magazine has the highest production cost per unit?

A	B	C	D	E
Utopia	*New Age*	*Sign of the Times*	*One People*	*Plan-it Power*

5 If 'margin' is defined as the difference between cover price and production cost per unit, what is *Green Earth*'s margin expressed as a percentage of production cost per unit?

A	B	C	D	E
£1.26	43.86%	54%	68.04%	198.4%

COMPARABLE MAGAZINE CIRCULATION FIGURES

For the last 12-month period

Magazine name	Average monthly circulation during given year	UK cover price
Utopia	210,478	£2.50
Green Earth	166,432	£1.80
New Age	1,129,407	£2.80
Natural Living	104,003	£1.15
Networker	78,024	£1.10
Sign of the Times	48,075	£4.20
One People	47,036	£3.15
Plan-it Power	38,143	£2.00

$$\text{Production cost per unit (£s)} = \frac{100{,}000}{\text{Average monthly circulation} + 20{,}000}$$

1 What is the price of *Utopia* expressed as a percentage of the price of the most expensive magazine?

Correct answer is A: 59.52% – The price of *Utopia* is £2.50. The price of the most expensive magazine *Sign of the Times* is £4.20

The percentage of the *Sign of the Times* price represented by *Utopia*'s cover price is therefore: 2.50 ÷ 4.20 × 100 = 59.52%

Therefore, the price of *Utopia*, expressed as a percentage of the price of *Sign of the Times* is **59.52%**

2 Assuming the whole circulation paid the full cover price, what was the average monthly income from sales during the last 12-month period for the three magazines with the highest circulations?

Correct answer is D: £1,329,370.70 – The 3 magazines with the highest circulations are *New Age*, *Utopia* and *Green Earth*. You first need to work out their individual average monthly incomes using their circulation figures and cover prices:
New Age: £1,129,407 × £2.80 = £3,162,339.60
Utopia: £210,478 × £2.50 = £526,195
Green Earth: £166,432 × £1.80 = £299,577.60

To calculate the average monthly income from sales of all 3 of these magazines you need to add these sub-totals together and divide by 3 (their number): £3,162,339.60 + £526,195 + £299,577.60 = £3,988,112.20. £3,988,112.20 ÷ 3 = £1,329,370.73

Therefore the average income from sales during the last 12-month period for the 3 magazines with the highest circulations was **£1,329,370.73**

3 If 5% of the cover price is paid to newsagents for each *Utopia* magazine sold, and *Utopia* is a monthly magazine, estimate from the given information how much was paid to newsagents in total over the last 12 months.

Correct answer is D: £315,717 – You calculated the average monthly income from *Utopia* sales based on its circulation and cover price in the last question; you need this as a starting point so you can calculate the value of the 5% paid to newsagents each month:
Utopia average monthly income: £210,478 × £2.50 = £526,195
5% of this income: £526,195 ÷ 100 × 5 = £26,309.75

You're asked for the total amount paid to newsagents over a year, so you need to multiply this by 12: £26,309.75 × 12 = 315,717

Therefore newsagents were paid **£315,717** based on 5% of *Utopia* sales over the last 12 months.

4 Which magazine has the highest production cost per unit?

Correct answer is E: *Plan-it Power* – You need to use the production cost per unit calculation provided in the numerical information for this question, but only for the magazines included in the multiple-choice options:
Utopia: 100,000 ÷ (210,478 + 20,000) = £0.43
New Age: 100,000 ÷ (1,129,407+ 20,000) = £0.09
Sign of the Times: 100,000 ÷ (48,075 + 20,000) = £1.47
One People: 100,000 ÷ (47,036 + 20,000) = £1.49
Plan-it Power: 100,000 ÷ (38,143 + 20,000) = **£1.72**

Therefore the magazine that has the highest production cost per unit is *Plan-it Power* with **£1.72** per unit.

A quicker way of answering this question would be to look at the production cost per unit calculation and see that the magazine with the lowest circulation would inevitably have the highest production unit cost, which is *Plan-it Power*.

5 If 'margin' is defined as the difference between cover price and production cost per unit, what is *Green Earth*'s margin expressed as a percentage of production cost per unit?

Correct answer is B: 42.86% – First you need to calculate the production cost per unit for Green Earth using the formula we used in the last question: *Green Earth*: 100,000 ÷ (166,432 + 20,000) = £0.54

The difference between cover price and production unit cost is £1.80 – £0.54 = £1.26

The percentage of this margin in relation to the production unit cost is therefore: £0.54 ÷ £1.26 × 100 = 42.86%

Therefore *Green Earth*'s margin expressed as a percentage of production cost per unit is **42.86%**

How did you do?

Score of 5	Score of 4	Score of 3	Score of 2	Score of 1
Great performance	Good performance	Average performance	Below average	Poor performance

UTOPIA MAGAZINE ADVERTISEMENT RATES

Colour or black and white	Size	Position	Front half of magazine	Price
Colour	Full page	Not specified	Not specified	£5,500
Colour	Full page	Facing editorial	Not specified	£5,950
Colour	Full page	Facing editorial	Front half	£7,300
Black and white	Full page	Not specified	Not specified	£4,000
Black and white	Full page	Facing editorial	Not specified	£4,500
Black and white	Full page	Facing editorial	Front half	£5,500
Colour	Half page	Not specified	Not specified	£3,200
Black and white	Half page	Not specified	Not specified	£2,550

Full page colour special positions

Inside front cover	£9,100
Inside back cover	£8,950
Outside back cover	£11,050
First double page spread	£18,500
Second double page spread	£16,300
Third double page spread	£15,500

Bleed: Add 10%

1 If a company buys 3 double page spreads, how much will it cost them?

A	B	C	D	E
£15,500	£16,767	£18,500	£50,300	£55,500

2 Environ Ltd plans to place 4 full page colour advertisements in Utopia magazine, 2 in no specified position and the other 2 on the outside back cover. What is the total cost, including 'bleed' for all 4 advertisements?

A	B	C	D	E
£22,000	£27,000	£30,500	£33,100	£36,410

3 If a company negotiates a 3% discount for buying a full page black and white advertisement facing editorial every month for a year, how much will it have saved (in £) at the end of the year?

A	B	C	D	E
£135	£165	£1,620	£1,980	£4,500

4 New World Power paid £120,120 for a run of 12 advertisements, each with the same format/position. Which format/position did it buy?

A	B	C	D	E
Full page, colour, Facing editorial, Front half	Full page, colour, Facing editorial, Not specified	Full page colour, Inside front cover with bleed	Full page colour, Inside back cover	Cannot say

5 *Utopia* sells 2 colour double page spreads with 'bleed' to a customer and gives them an additional half page colour without charge. What discount is this equivalent to, expressed as a percentage of the cost of the 2 colour double page spreads?

A	B	C	D	E
7.86%	8.36%	8.64%	9.20%	Cannot say

UTOPIA MAGAZINE ADVERTISEMENT RATES

Colour or black and white	Size	Position	Front half of magazine	Price
Colour	Full page	Not specified	Not specified	£5,500
Colour	Full page	Facing editorial	Not specified	£5,950
Colour	Full page	Facing editorial	Front half	£7,300
Black and white	Full page	Not specified	Not specified	£4,000
Black and white	Full page	Facing editorial	Not specified	£4,500
Black and white	Full page	Facing editorial	Front half	£5,500
Colour	Half page	Not specified	Not specified	£3,200
Black and white	Half page	Not specified	Not specified	£2,550

Full page colour special positions

Inside front cover	£9,100
Inside back cover	£8,950
Outside back cover	£11,050
First double page spread	£18,500
Second double page spread	£16,300
Third double page spread	£15,500

Bleed: Add 10%

1   If a company buys 3 double page spreads, how much will it cost them?

Correct answer is D: £50,300 – You need to refer to the second table in the numerical information and interpret it correctly – do not simply multiply the price of the First double page spread by 3. Instead you need to add together these values: First double page spread + Second double page spread + Third double page spread: £18,500 + £16,300 + £15,500 = £50,300

Therefore, if a company buys 3 double page spreads, it will pay **£50,300**

2

Environ Ltd plans to place 4 full page colour advertisements in Utopia magazine, 2 in no specified position and the other 2 on the outside back cover. What is the total cost, including 'bleed' for all 4 advertisements?

Correct answer is E: £36,410 – The first step is to work out the costs for each of the individual adverts:
2 × full page colour in no specified position: £5,500 × 2 = £11,000
2 × full page colour on the outside back cover: £11,050 × 2 = £22,100

The total for these 4 adverts will therefore be £11,000 + £22,100 = £33,100

But don't forget the bleed! This costs an additional 10%: £33,100 × 110 ÷ 100 = £36,410

Therefore the cost to Environ Ltd for the 4 full page colour advertisements with bleed will be **£36,410**

3

If a company negotiates a 3% discount for buying a full page black and white advertisement facing editorial every month for a year, how much will it have saved (in £) at the end of the year?

Correct answer is C: £1,620 – Start off by calculating what the 3% discount will be on a full page black and white advertisement facing editorial:
3% discount on a full page black and white advertisement facing editorial: 4,500 ÷ 100 × 3 = 135

£135 is the monthly saving so by the end of the year the saving will be 12 times greater than this: 135 × 12 = 1,620

Therefore the company saves **£1,620** in a year by negotiating a 3% discount for buying a full page black and white advertisement facing editorial every month.

4

New World Power paid £120,120 for a run of 12 advertisements, each with the same format/position. Which format/position did it buy?

Correct answer is C: Full page colour, Inside front cover with bleed – You need to calculate the prices of the adverts included in the multiple-choice options:
Full page, colour, Facing editorial, Front half: £7,300 × 12 = £87,600
Full page, colour, Facing editorial, Not specified: £5,950 × 12 = £71,400
Full page colour, Inside front cover with bleed: £9,100 × 12 = 109,200 + 10% = £120,120
Full page colour, Inside back cover: £8,950 × 12 = £107,400

Therefore New World Power bought 12 **Full page colour, Inside front cover adverts with bleed** for £120,120

5	*Utopia* sells 2 colour double page spreads with 'bleed' to a customer and gives them an additional half page colour without charge. What discount is this equivalent to, expressed as a percentage of the cost of the 2 colour double page spreads?

Correct answer is B: 8.36%. – First you need to calculate the usual cost of 2 colour double page spreads with 'bleed': First double page spread (£18,500) + Second double page spread (£16,300) = £34,800

This price with bleed is £34,800 + £3,480 = £38,280
The additional half page colour advert is worth £3,200

You can now calculate what the value of this free advert is as a percentage of the price paid for the other 2 adverts: £3,200 ÷ £38,280 × 100 = 8.36%

Therefore this is equivalent to a discount of **8.36%** of the cost of the 2 colour double page spreads.

How did you do?

Score of 5	Score of 4	Score of 3	Score of 2	Score of 1
Great performance	Good performance	Average performance	Below average	Poor performance

SO HOW DID YOU DO?

These practice tests are too short individually to make an accurate measurement of your numerical reasoning ability, but if you add your scores on all 10 practice tests we can provide some advice. Compare your total score with the table below to gauge your performance.

Total score	What it means	Advice for further development
50	Excellent performance	It looks like you're pretty much there in terms of developing your personal best score. Don't rest on your laurels though. Make sure that you take the time to read the remaining chapters of this book; they contain information that will help you maintain this level of performance. Also, remember to keep exercising your raw numerical ability through activities such as performing daily calculations at work, solving number puzzles and reconciling your current account each month. This way you can keep your ability at its strongest until your next test.
40–49	Good performance	This level of performance should see you through most numerical tests, but if you feel that it does not reflect your potential personal best score then you might want to do some more practice. Review the advice in Chapter 2 about test strategies and practising and then have another go at the practice questions.
30–39	Average performance	Numerical critical reasoning tests are the hardest type of numerical assessment you are likely to take, so an average score is still a positive result. However, you could benefit from doing some more work to practise and pass numerical tests at this level more effortlessly. Review the advice in Chapter 2 about test strategies and practising. You can also review your wrong answers to the practice questions to identify any trends in where you go wrong. It might be that you're making simple errors in your calculations or perhaps being too easily taken in by the multiple-choice distracters. Whatever the case, spend some time reviewing where you went wrong and then attempt the practice questions again.

Total score	What it means	Advice for further development
20–29	Below average performance	These are high-level tests so you might want to polish your numerical ability by correcting any shortfalls in your numerical reasoning abilities by revisiting your work in Chapter 4. Once you are happy that your performance at those mid-level tests reflects your true potential try these numerical critical reasoning tests again. To begin with do not apply the time limits when you attempt these high-level tests – see if that makes a difference. If time is a factor then you should identify which areas of your test-taking strategy might be letting you down. Spend some time developing your raw numerical ability through activities such as mental arithmetic, solving number puzzles and taking responsibility for budgeting at work or home alongside further practice from this book.
0–19	Low performance	

Remember that these 10 practice tests are pitched at the highest level of numerical skills you are likely to require in the workplace. If your personal best is not within the top range of the results then it might simply be the case that your strengths lie elsewhere – perhaps within another ability like your verbal reasoning. You can offset a deficit within one area of your performance by focusing on your strengths. If you are going to complete a numerical test it is likely that you will be asked to complete a different test alongside it. You can maximise your overall performance by developing all of the abilities you are tested on. Consider working through other books in the Practise & Pass series to help you build on all your strengths.

CHAPTER 6
ONLINE NUMERICAL TESTING

The traditional format for numerical ability testing is using paper and pencil materials and a test administration session held under supervised conditions, typically with groups of people. Over the course of the past 10 years this has changed, with employers using the internet to deliver tests to their candidates. Today, if you are asked to take a test for an employer, it is highly likely that you will complete it online. This change in testing format has implications for your personal best score, which are explored in this chapter.

HOW IT WORKS

Online testing is most commonly used by large employers within high volume recruitment processes. The greater the number of people who apply for a vacancy, the greater the

---

INSIDER INFO

FREE ONLINE TESTS

A quick internet search for free online numerical tests will yield millions of results – but take care, because the vast majority will be of questionable quality compared with the tests used by employers to select people for recruitment and development.

Reputable online ability tests are built to high standards of accuracy and have been subjected to research to ensure that they are fit for purpose. Test publishers in the UK submit their tests for review by the British Psychological Society to enable test-users to make informed decisions about which tests to use based on the technical qualities of the test.

If you are looking online for a practice numerical test you are encouraged to restrict your search to UK-based test publishers. This way you can feel confident that any practice questions you complete are representative of the format, style and quality of real psychometric tests of ability.

For a list of online practice tests, see Chapter 9.

likelihood that the employer will adopt online testing instead of paper and pencil testing. However, employers do use online testing for a range of purposes unconnected with recruitment. Many of our clients ask their training delegates to complete an online test before they attend a leadership development programme, therefore saving time during the course itself, which would otherwise be taken up with a testing session. This saved time can be devoted to exploring the results of the test.

There are two common approaches to online testing:

▶ **Supervised online testing**. This approach is not all that different from traditional paper and pencil testing. To complete the numerical test you attend the employer's testing centre (their offices or other location) and complete the test on a computer in the presence of a test administrator, who oversees the testing session. The only difference is that you are completing the test on a computer rather than with paper and pencil materials. If you are asked to complete an online test under supervision then there is little difference with traditional testing in terms of the approach you should take to achieving your personal best scores.

▶ **Unsupervised remote online testing**. This approach is the one most favoured by employers because it benefits them in terms of time and cost. Instead of inviting candidates to complete the test under supervision, the test administrator sends candidates a link and password for the online numerical test. Candidates complete the online test in their own time in a location of their choosing. This approach removes the requirement for the employer to provide a testing location or people to oversee the test session. If you are asked to complete an online numerical test remotely and unsupervised you may need to adapt your approach in order to maintain the level of performance you'd expect to achieve in a traditional testing session.

The first step is to understand the differences between traditional testing and unsupervised online testing.

WHAT ARE THE DIFFERENCES?

The obvious difference between an online test and a traditional paper and pencil test is the format. You complete the test using a PC connected to the internet, which displays the test instructions, the numerical information and the numerical questions. You enter your answers on the screen rather than on a separate answer sheet.

The unsupervised element of online numerical testing means that you will take the test alone, at a time and place of your choosing, with no test administrator watching you. For many people this is a relief from the nerve racking environment of the traditional test session.

You are required to provide the testing materials, that is you need to have an internet-connected computer to complete the test. This is an aspect of the cost of testing that the employer is passing on to you. For many people this presents no problem since they have access to the internet on a PC at home or work; but it can cause difficulties for those who do not.

While you are expected to provide the testing materials the remote nature of the online numerical test does offer you savings as well in terms of removing the need for you to travel to the employer's place of business to be tested. Our clients who were the earliest adopters of online testing did so because previously they needed to fly candidates into their UK office from all over Europe and Asia to be tested. Remote testing had clear benefits for the employer and the candidates.

INSIDER INFO

ONLINE RECRUITMENT – THE FUTURE OF JOB APPLICATIONS?

Over the past 10 years online recruitment has replaced high street recruitment, where you visit a job agency or look in the local paper, as the dominant approach for advertising and accepting applications for jobs.

Websites such as Monster and Totaljobs have become the first stop for many employers and candidates in the recruitment market. The busiest time for a jobs website tends to be Monday afternoon around 2 p.m. – this is when the peak numbers of job searches happen. Perhaps when most people feel most depressed about their current job following the weekend?

Online tests have grown alongside the job websites and employers' careers websites. Often you will be passed seamlessly from a website where you completed an online application form or uploaded your CV to an online test site. Once you have completed the test the employer will assess your scores along with the information you submitted with your application to make a decision about whether to invite you to interview.

The benefits for employers and candidates tend to be in terms of streamlining and efficiency. It takes less time for employers to process online applications and test results than traditional paper-based approaches. Candidates can be kept up to date with the progress of their applications by email and text messaging.

There is still debate among some business psychologists about whether tests should be used online but ultimately it is the pragmatism of employers and many candidates that is driving online recruitment and testing – the time savings and convenience offered by the new technology is causing recruitment and testing processes to slowly change.

Your results are calculated by the online testing system rather than the test administrator, who traditionally would have to calculate results by hand. The benefit for you as the test taker is the removal of any possible scoring error that might adversely affect your score.

When the employer is using the online test for recruitment they will often use your numerical test results, alongside any other information you have provided such as a CV or application form, to make a pre-selection decision. This means the employer sifts out any applicants whose numerical test score and supporting information indicate low suitability for the job. Only candidates with the right level of ability combined with the right experience will be invited to interview. Many candidates worry about this high-stakes nature of the online test.

When employers use an online test as part of a training programme their intention is typically to explore the results during the subsequent course to aid the individual's development and self-awareness. Online testing in this context feels less threatening since it is lower stakes.

WHAT'S THE BEST APPROACH TO ACHIEVING YOUR ONLINE PERSONAL BEST?

Here's some advice you should follow when taking an online numerical test remotely:

Your testing environment

You can choose the location where you will take the test as long it has an internet-connected computer. However, to perform at your best bear in mind the following points.

▶ **Aim to make your personal testing session as close in conditions to a traditional, supervised session as possible**. This will enable you to perform at your maximum level.

INSIDER INFO

DO PAPER AND PENCIL TEST RESULTS AGREE WITH ONLINE RESULTS?

The short answer is 'broadly yes'. When we correlate test scores gathered using traditional paper and pencil testing under supervised conditions with online numerical test scores gathered using remote testing we see a strong degree of agreement between the two sets of scores. In other words, the results we see from online tests tend to correlate well with traditional test scores.

There is a bit more going on within these findings, however. When we examine individual people's test scores from the two different types of test administration we see that although most people score the same on both tests, the scores of some candidates do change. Some people's numerical ability scores **decrease** when we test them under supervised conditions and compare this with their initial online performance. But other people's scores actually **increase** when they are tested under supervised conditions compared with their online results. The important thing to note is that this happens whenever we test a group of people more than once, regardless of the type of test (online, or paper and pencil) or type of test administration format (supervised, unsupervised, remote) and compare their results. The results we see tend to be the same for most people but a number of candidates will see an increase or decrease in their second set of test scores. This is due to practice effects. On the second test session people vary in their approach because of practice: some are more relaxed; some more complacent; some are less anxious; some have developed better strategies. These factors lie beneath the changes we can see in some people's test scores.

The good news for you is that this book is designed to help you capitalise on practice effects by developing your test-taking prowess to increase your next test score, regardless of whether it is online or on paper. You can also relax about online tests – they tend to give the same numerical ability results as their traditional, paper and pencil, supervised alternatives.

▶ **Choose a quiet, relaxed environment** free from distractions – consider putting a 'do not disturb' sign on your door. Switch off your phone.

▶ **Complete the test in a single sitting** – do not take a break halfway through a test; this will break your concentration. Some online numerical tests will not let you complete a test, by timing out if you left the system prematurely.

▶ **Give yourself sufficient time** so that you do not have to rush. Most online tests have a strict time limit – make sure you make the most of the time allowed.

▶ **Focus all of your attention on completing the test.** If the doorbell or phone rings during the test, ignore it! If it's important the caller will try again later.

Test-taking strategies

As stated above, the test-taking strategies you have developed using this book can all be transferred to remote online numerical testing. There are a few additional points, however, specific to online tests.

▶ **Read all onscreen instructions very carefully.** In a traditional supervised test session the administrator will read through the test instructions with you to ensure you understand how to proceed. With remote testing there is no administrator so it is up to you to ensure you are clear about what you have to do before you start the test itself. There is no time limit for reading the instructions; so take your time.

▶ **There can be no going back!** Online tests often vary in their format from paper and pencil. Do not make the assumption that the instructions for completing the test are going to be the same as those for a paper and pencil test. One important example is the use of the 'back' button in your web browser; many online tests prevent you going back to pages you have already completed to change your answers. This is different from a paper and pencil test where reviewing all your answers once you've attempted all the questions is excellent test-taking strategy. With an online numerical test you should take some time to review your answers before you move on to the next

INSIDER INFO

WHY IS STANDARDISATION IMPORTANT?

In traditional testing, the tests are taken under standardised conditions. This means that the test administrator has to follow the administration instructions for the test very carefully to ensure that the test session is identical in format, approach and timings to every other test session that is conducted for the same numerical test. One of the reasons for this standardised approach is to ensure that every candidate receives an equivalent experience to every other candidate, thereby making the assessment fairer. But there are important psychometric reasons for standardising the testing conditions as well. Test developers need to ensure that test sessions are as close to identical to each other as possible to maximise the effectiveness and accuracy of the test and the scoring mechanism. In doing so, test developers reduce the number of errors within the testing process that might affect your score (upwards or downwards).

When you are taking a test online it is much harder for the test developers to guarantee the same standardisation of testing conditions because the test does not take place under supervision. This places demands on the design of the test itself to ensure that it will still measure numerical ability accurately and consistently.

It will benefit your test score if you strive to standardise the conditions of your own online test. This means following the advice above and the instructions for the test very closely.

page – you may not be able to go back when you reach the end of the test.

▶ **Untimed tests**. Some online numerical ability tests do not have time limits. These tests are open-ended and allow you to work at your own pace. In the absence of a time limit the test will often use other techniques to get a good measurement of your ability, such as questions that become increasingly difficult as you progress through the test. If you complete an untimed test you should still follow the usual test-taking strategy of working quickly but accurately because the length of time you take to complete the test is very likely to be recorded. The time you take to finish may be a factor that is considered alongside your actual numerical ability score. Do not go off to make a cup of tea halfway through a test without a time limit – the clock will still be ticking in the background!

▶ **Online anxieties**. Most people find remote testing less worrying than supervised testing because of the more relaxed environment and absence of a test administrator. Some people do find it causes them anxiety, however. The test-taking strategies for reducing nerves during a traditional test session still transfer to online testing. Do not worry about a pass/fail score if the numerical test is being used as part of a sifting process prior to the interview – the sifting decision will not be based on your test score alone. Any additional information you have provided, such as your CV, also inform the decision.

Technical advice

You do not need a super-computer to take an online numerical test, and a well-designed web-based testing system will not discriminate against you if your PC is getting on a bit. However, there are some points relating to the technology you use that will help you to achieve your personal best:

▶ **Use an up-to-date web browser**. Good online testing systems are designed to run on any computer, even an Apple computer. The important factor is not the type of computer or its age, but the type of browser you are using. Most people use Microsoft's Internet Explorer because this comes pre-installed on all PCs. Online numerical tests are typically designed to run best on versions of this browser that are newer than version 6. You can check the version of your browser by clicking on 'Help' in its menu bar and selecting 'About Internet Explorer'. Updating to a more recent version is free. If you are using an Apple computer or a different browser on your PC (such as Firefox or Google's Chrome) then the same advice applies – update to the most recent version. Using older versions of browsers will not stop an online numerical ability test from working but they can cause odd display issues and affect the layout of the test as it appears on your screen – this could potentially affect your score.

▶ **Web-browser settings**. You may experience odd issues if you have custom or unusual settings switched on in your browser. Before taking an online test go to the Options menu and make sure that your browser settings are switched to default. Other settings may cause problems and affect your score.

▶ **Web connection speed**. You do not need a super-fast connection to take an online test, but a broadband connection will be preferable to a dial-up connection. A slower connection will not affect your score but the frustration of waiting for pages to display might do. Try to use a computer that has a broadband connection.

▶ **No access to a computer**. Not everyone has a computer at home or one they can use at work (especially when applying for alternative employment). The best advice is to use a computer at your local library, where internet access is generally free, or a friend, family member or colleague's computer or even an internet café.

If you definitely can't get access to a computer it is the responsibility of the employer who has asked you complete the test to provide alternative arrangements. Sometimes this is a paper and pencil version or an appointment to visit the employer and complete the online test at their office.

▶ **No web or computer experience**. Not everyone uses a computer or browses the internet. If you do not feel confident taking a test online then the best advice is to contact the employer and ask for an alternative test, such as a paper and pencil test. Although it is the test-user's responsibility to ensure all candidates have access to the testing, you should bear in mind that employers will only choose to use an online test if the job itself requires people to use a computer. If you lack confidence or experience with using computers or the internet you may be disadvantaged in terms of your suitability for a job that requires you to use a computer. You could consider developing your computer literacy or alternative jobs that do not require any IT skills.

▶ **Technical difficulties**. Broadband connections tend to be very robust, but if you are using dial-up there is a possibility that you might lose your connection when you are midway through an assessment. If you experience any technical difficulties when completing the online numerical test, such as a dropped web connection or the computer freezing, don't panic! The best advice is to switch your computer off and then switch it back on again. If you leave midway through a timed numerical test you may need to request a reset of the time limit from the individual responsible for the assessment process. You will be guided through this process when you try to resume the assessment. When you re-enter the test system it will either let you pick up where you left off or invite you to contact a helpdesk. Don't panic – your score will be unaffected. If you are worried though, you can ask the employer for another opportunity to take the test, perhaps at their premises.

CHEATS NEVER PROSPER

The temptation to cheat on a remote online numerical test is understandable; a lot can ride on the results and there's no test administrator to keep an eye on what you're up to. The potential for candidate cheating is a source of anxiety for employers who use online tests for remote and high-stakes assessment. Below is a list of the most common forms of cheating at an online numerical test.

▶ **Ask a friend to take the test for you**. You might know someone who you feel has higher levels of numerical ability than you, whom you could ask to take the test for you. Alternatively you could ask the friend to sit with you while you complete the test or just phone your friend if you encounter any difficult questions. If you use this approach make sure that your friend's numerical ability really is better than your own.

▶ **Find the correct answers for the test**. If you know the precise name of the numerical test you will take you could try a web search for the correct answers. Websites do exist that enable people to share what they believe are the correct answers to widely used online tests; these are sometimes called 'cheat-sheets'. You simply enter these correct multiple-choice answers when you complete the numerical test yourself.

▶ **Take the numerical test more than once**. You can complete all the questions in the test as a practice go and then re-apply to the testing system to have another go.

Should you decide to employ any of these methods to improve your test score you should be warned that test developers

have become very adept at preventing and detecting efforts to cheat – you will be found out! If you decide to cheat, here's what you're up against.

▶ **Randomised questions**. Most online tests include a degree of randomisation that means the numerical information and questions you are presented with are drawn from a large pool. This means that if you take a numerical test more than once it is unlikely that you will see the same questions. If you try to find the correct answers for an online test the cheat-sheet you download is unlikely to match the randomised set of questions and answers you are presented with.

▶ **Secure scoring**. Where cheat-sheets do exist they are based on other people's beliefs about what the correct answers to a test actually are. The correct answers to a test are never made public, so cheat sheets may not in fact be an accurate description of the right answers for a test. Widely used numerical tests are frequently updated to stay one step ahead. The questions are changed to minimise any over-familiarity among candidates.

▶ **Validation of identity**. Nearly all online tests require a pass code for entry, which is typically only good for one go. This means that if you complete the numerical test once as a practice, you will not be able to get back into the test system to take it again. Online tests take further measures to validate your identity – to ensure it is you who is taking the test and not a friend on your behalf.

▶ **Scores can go down as well as up**. If you do manage to fool an online numerical testing system to let you have more than one attempt you should be aware that not everyone's score increases on subsequent attempts. Research into the effects of multiple completion of the same test indicates that for many candidates, scores can go down or simply stay the same.

▶ **Verification of results**. If your numerical ability is tested remotely online, any employer who is following simple best practice guidelines will re-test your numerical ability under supervision to verify the results. This re-testing can take the form of a paper and pencil test, a supervised online test when you attend interview or a different assessment method entirely. It is not unusual for employers to use a written exercise alongside the interview to verify your level of numerical ability. If your online test score was inflated through cheating you will be found out!

If you try to cheat at an online test all or one of these measures will find you out!

Test developers have several other measures that are designed to deter, resist and detect efforts to cheat so on balance our advice would be don't try your luck. If you are fortunate enough to cheat at a numerical test and get the job, remember that your new employer will expect to see you demonstrate the same level of numerical ability in your performance at work as your online test score.

All methods of testing, selection and assessment are open to abuse by candidates who want to get the job, not least the interview. Ultimately it is in your best interests to be honest and avoid ending up in a role to which you are not suited in terms of your numerical ability.

CHAPTER 7
WHAT HAPPENS AFTER TAKING THE TEST?

What a weight off your mind – you've completed the numerical test and now you can relax! But just because the test is over doesn't mean that you can forget about it. There are opportunities to develop your personal best numerical test score further still if you make the right use of any results that are communicated to you.

In order to develop your personal best score based on your numerical test results you need to understand a little more about test scores, what they mean and how to maximise the benefit you gain from any feedback about your performance.

HOW TEST RESULTS ARE USED

Test scores should never be used in isolation. Whether the numerical test was used as part of a recruitment or development process the results should always be placed in the context of other information about you. This might mean combining your numerical test results with scores from another test, such as a verbal test, or with a different type of assessment entirely such as a personality questionnaire or interview.

When placed into context with other assessments of your ability and/or personality your numerical test score can tell you something about how your ability in this area integrates with your other strengths or weaknesses. This overall interpretation enables potential employers to make a more informed selection decision about you, or if the test formed part of a development process it will enable you to develop greater self-awareness.

The test user should offer you feedback about your numerical test results but this will often be in the context of a wider set of results. Within this feedback you need to understand what the results mean if you are to interpret and learn from your numerical test score.

INSIDER INFO

PERCENTILE SCORING

The majority of numerical test results are based on a scoring system called percentiles. A percentile score tells you what proportion of people who've taken the test before you have scored better than (or worse than!).

The comparison group of people who have taken the test before is called a 'norm group'. Your performance on the numerical test is compared with the norm group and the resulting percentile score describes your numerical ability in relation to this group.

So, if you score on the 75th percentile you scored better than three-quarters of the norm group – or you are in the top 25%. If you scored on the 50th percentile you are better than the bottom half of the norm group, but worse then the top half (the 50th percentile is the absolute average in percentile terms but most test users interpret any percentile score between the 35th and 64th as lying in the average range of ability).

It is important not to confuse percentile scores with percentage scores. A percentile score tells you how well you performed in relation to others; it does not tell you what percentage of the test questions you scored correctly.

The composition of the norm group your performance is being compared with is an important piece of information in terms of understanding your numerical ability percentile score. You should be compared with a group of people who are similar to you in terms of educational level, work experience and the type of job for which you've applied. Ideally the norm group will be made up of people who have taken the test for the same employer as you have and for the same purpose or job.

A trained test-user will be able to clearly explain your percentile score and the nature of the norm group with which your scores have been compared during the feedback process.

WHAT THE RESULTS MEAN

Your numerical test score is unlikely to be simply the number of questions you answered correctly. Your results will be calculated on the basis of how your performance compares with people who have taken the same numerical test before.

This comparison group will be made up of people whose background, educational level or job is broadly representative of candidates who are taking the test for the employer who has asked you to complete the assessment. Typically the group will comprise over 100 people whose performance on the test will range from excellent to poor.

This comparison enables the test-user to understand how strong your numerical ability is in relation to other people. The comparison consists of a calculation that produces your numerical ability result.

Typically this result is framed in terms of how well you performed compared to other people. It is unlikely that you will be given a numerical score or grade as you would with an educational test. Instead the test user will describe your result in terms of how close it is to the average score. Here are some examples of how real numerical test scores are fed back to candidates.

'Your numerical ability is well above average compared to a group of managers who have taken the test before.'

'Your numerical test result lies within the bottom 20% of other graduates who applied to the same role.'

'Your numerical ability lies within the average range for call centre workers; it is typical of other people working in call centre roles.'

The test-user will not feed back the number of questions you answered correctly because this is not the way that the results from modern psychometric tests of numerical ability are scored and interpreted. In the case of online tests, where the scoring is conducted automatically by the computer, the test-user is unlikely to know how many questions you answered right or wrong. Your results will be presented in terms of the comparison to other test takers.

GETTING THE MOST FROM TEST FEEDBACK

Only trained test-users should feed back your results; this ensures that the feedback you receive is based on expert knowledge of the test and what its results mean. While the numerical test score itself is generally the piece of information with which candidates are most preoccupied, the broader feedback process has much greater potential for helping you improve on your numerical test performance in future.

Numerical test feedback is an excellent opportunity to discover the detail identified about your numerical ability by the test. Based on my own and the experiences of other occupational psychologists and expert test-users, here are some points to bear in mind when receiving test feedback.

▶ **Be honest**. The test-user providing your feedback may prompt you to describe your reactions to the test and how you feel your results will look. It will help improve the quality of the feedback discussion if you are honest with yourself and the feedback provider about your experiences of the test and your prediction of your performance. Sharing your experience of the test will enable the test-user to shape the feedback to help you identify the aspects of your performance that could be improved next time. If for example you found the numerical test difficult or confusing then tell the test-user. This is useful information that they can use to help you develop your approach in future.

▶ **Don't be defensive**. The test user will feed back your results using the kind of language described above; how your performance compares to other candidates. This means that the test user may use terms such as 'below average', or 'bottom 15%' to describe your results. Receiving feedback can be an emotional process and you should put any defensive reactions to one side. Avoid arguing about the outcome – remember that psychometric numerical ability tests have very high levels of reliability; their measurements tend to be very accurate. If your test result is a surprise (perhaps lower than you expected) then you should use the feedback discussion to identify the reasons that may have contributed to the result – this might give you a clue about how to improve your performance next time.

▶ **Ask follow-up questions**. You can ask the feedback provider to describe what your numerical ability test score means in relation to the job. If the test was used for recruitment purposes the test user should be able to describe what elements of the role would be affected by your test results. Where the numerical test is being used as part of a development process you should ask the test-user to identify the implications that the result has for your future development – do the results suggest a development need for your numerical ability?

▶ **Ask for advice**. The feedback provider should be able to provide you with advice for improving your future test performance that is specific to the test you have taken. You can prompt the test user to identify actions you could take to improve your test-taking strategy, your knowledge of the specific numerical test used and your underlying numerical ability.

Test users are encouraged by test developers and the British Psychological Society to always offer feedback to test takers.

INSIDER INFO

BEST PRACTICE IN ABILITY TEST FEEDBACK

Test users who have qualified with the British Psychological Society to use ability tests are trained to:

- ▸ Reiterate **why the test was used**
- ▸ Describe the **role of the test score** in any selection decision
- ▸ Provide a short but clear **description of the test**
- ▸ Explain issues of **confidentiality and data storage** regarding your test results
- ▸ Ensure that feedback is an **interactive** process in which the test user discusses the findings with the test-user
- ▸ Describe the results in **clear, lay terms**
- ▸ **Relate the results to the position** for which you've applied
- ▸ Ask for your **comments and reactions**
- ▸ Provide you with an opportunity to **ask questions** about your test results

As a test taker you should expect the feedback you receive to reflect this best practice approach.

Not every employer adopts this best practice approach so it is always worth confirming whether test feedback will be available before taking the test. Chapter 8 explores rights such as these that you should expect to be fulfilled as a test taker.

CHAPTER 8
YOUR TESTING RIGHTS

P sychometric tests of ability should only be used by trained specialists. This might be an occupational psychologist or someone who holds recognised occupational testing qualifications. Trained test users should stick to best practice principles of testing that ensure that your testing rights are met.

As with any technology, tests can be abused by people who have not been properly trained or who simply do not conform to best practice. As a test taker you should be aware of your testing rights and testing best practice more generally. When a test is misused or best practice ignored the measurement of your ability is likely to be distorted or misinterpreted. It is very difficult to achieve your personal best score under such circumstances.

HOW YOU SHOULD BE TREATED

A key principle in testing is 'informed consent'. As a test taker you have a right to be fully informed about the nature of the assessment process before you give your consent to be tested. This means that the test user must provide you with these key pieces of information before you consent to take the test.

- **What the test measures** – in this case they would tell you that it is a test of your numerical ability.
- **What the results will be used for** – the test user must be clear about how the test results will be used within their recruitment or development process. The results must never be used for purposes other than those to which you have consented.
- **Who will see your results** – the test user must provide clear assurances about the boundaries of confidentiality around your test results. This means being clear about who will see your test scores.
- **How long your results will be stored** – you need to be told how long the employer will hold on to your results. Six to 12 months is a typical length of time.

▶ **Provision of feedback** – The employer must be clear about whether or not they will provide feedback. Best practice is to offer feedback to all candidates, but many employers feel unable to do so because of high volumes of applicants or a lack of resources. Many employers therefore provide feedback on request to unsuccessful candidates and encourage feedback to be taken up by successful ones. Other employers offer no feedback at all. Your fundamental right as a test taker is to have a clear understanding about what feedback, if any, will be available once you have completed your numerical test.

Usually this information should be provided when the employer contacts you to invite you to complete the test. The letter will either ask you to confirm that you consent to be tested or require you to contact the employer if there are any conditions you are not happy with. Hand-in-hand with this right to informed consent you have certain responsibilities as a test taker. If there are any factors that might affect your test performance you must let the test user know so they can make any appropriate adjustments. Common candidate needs of this kind are impaired vision, dyslexia or language difficulties. If you feel that your test performance may be adversely affected by factors such as these, it is your responsibility to bring them to the attention of the test-user before the testing session. Only this way can the test user take actions that will ensure you are able to demonstrate your personal best.

As a test taker you should also expect to not be adversely affected by the experience of being tested. This is called the 'principle of self-regard', which states that test takers should feel as good about themselves at the end of a test session as they did at the beginning. Test users have a responsibility to ensure that testing is appropriate, related to the job and not excessive.

If you experience a test session that you feel breaks the principle of self-regard you might consider whether the employer is an attractive one for you to work for.

TESTING BEST PRACTICE

The British Psychological Society (BPS) has established a set of standards that test users should adhere to in order to ensure that tests are used in a fair, effective and ethical way. Test users must hold the BPS qualification in occupational testing called Level A in order to use ability tests.

When you are invited to a testing session you can check whether the employer holds this Level A training by contacting the BPS, which keeps a register of qualified test users (contact details are given in Chapter 9). Individuals who are not Level A qualified are unlikely to adhere to the levels of best practice that you should expect as a test taker. The testing may not be fair, effective or ethical.

Level A trained test users will also understand the nature and quality of different numerical tests and select the one that is most relevant and appropriate for their testing purposes. Trained test users know which tests to avoid on the basis of quality or reliability.

The BPS standards were established to protect test takers from being subjected to unfair, inappropriate and unethical test use. You can find out more about the standards by visiting the BPS website.

CHAPTER 9
FURTHER RESOURCES

There is more help and support out there for test takers. Once you have worked through this book you may find it helpful to continue your preparation using the resources below.

PRACTICE RESOURCES

For practice questions provided by test publishers, visit:

▶ http://criterionpartnership.co.uk/psychometrics_help
▶ www.shldirect.com/practice_tests.html

TEST SOPHISTICATION RESOURCES

For advice on how to approach tests, see:

▶ http://criterionpartnership.co.uk/takers_advice
▶ www.shlgroup.com
▶ www.savilleconsulting.com/products/aptitude_preparationguides.aspx
▶ www.psychtesting.org.uk/ptc/roles$/the-public.cfm

TEST KNOWLEDGE AND GOOD PRACTICE RESOURCES

▶ The British Psychological Society (BPS) Psychological Testing Centre is the body that promotes test standards. A number of useful resources can be found on its website, including the Level A standards, test takers guide and the BPS register of qualified test users.
▶ The International Test Commission (ITC) has established a set of standards for online testing: International guidelines on computer-based and Internet delivered tests, which you can find here: http://www.intestcom.org/guidelines/index.php